Review of the Future of the U.S. Aerospace Infrastructure and Aerospace Engineering Disciplines to Meet the Needs of the Air Force and the Department of Defense

Committee on the Future of the U.S. Aerospace
Infrastructure and Aerospace Engineering Disciplines to
Meet the Needs of the Air Force and the Department of Defense

Air Force Science and Technology Board
Division on Engineering and Physical Sciences
National Research Council

NATIONAL ACADEMY PRESS
Washington, D.C.

NATIONAL ACADEMY PRESS **2101 Constitution Avenue, N.W.** **Washington, DC 20418**

NOTICE: The project that is the subject of this report was approved by the Governing Board of the National Research Council, whose members are drawn from the councils of the National Academy of Sciences, the National Academy of Engineering, and the Institute of Medicine. The members of the committee responsible for the report were chosen for their special competences and with regard for appropriate balance.

This is a report of work sponsored by the Air Force Office of Scientific Research under grant number F49620-99-1-0338 between the Department of Defense and the National Academy of Sciences. Any opinions, findings, conclusions, or recommendations expressed in this publication are those of the author(s) and should not be interpreted as necessarily representing the official policies or endorsements, either expressed or implied, of the Air Force Office of Scientific Research or the U.S. Government.

International Standard Book Number 0-309-07606-4

Copies are available from:

National Academy Press
Box 285
2101 Constitution Ave., N.W.
Washington, DC 20055
800-624-6242 or 202-334-3313 (in the Washington metropolitan area)
http://www.nap.edu

Printed in the United States of America

THE NATIONAL ACADEMIES

National Academy of Sciences
National Academy of Engineering
Institute of Medicine
National Research Council

The **National Academy of Sciences** is a private, nonprofit, self-perpetuating society of distinguished scholars engaged in scientific and engineering research, dedicated to the furtherance of science and technology and to their use for the general welfare. Upon the authority of the charter granted to it by the Congress in 1863, the Academy has a mandate that requires it to advise the federal government on scientific and technical matters. Dr. Bruce M. Alberts is president of the National Academy of Sciences.

The **National Academy of Engineering** was established in 1964, under the charter of the National Academy of Sciences, as a parallel organization of outstanding engineers. It is autonomous in its administration and in the selection of its members, sharing with the National Academy of Sciences the responsibility for advising the federal government. The National Academy of Engineering also sponsors engineering programs aimed at meeting national needs, encourages education and research, and recognizes the superior achievements of engineers. Dr. Wm. A. Wulf is president of the National Academy of Engineering.

The **Institute of Medicine** was established in 1970 by the National Academy of Sciences to secure the services of eminent members of appropriate professions in the examination of policy matters pertaining to the health of the public. The Institute acts under the responsibility given to the National Academy of Sciences by its congressional charter to be an adviser to the federal government and, upon its own initiative, to identify issues of medical care, research, and education. Dr. Kenneth I. Shine is president of the Institute of Medicine.

The **National Research Council** was organized by the National Academy of Sciences in 1916 to associate the broad community of science and technology with the Academy's purposes of furthering knowledge and advising the federal government. Functioning in accordance with general policies determined by the Academy, the Council has become the principal operating agency of both the National Academy of Sciences and the National Academy of Engineering in providing services to the government, the public, and the scientific and engineering communities. The Council is administered jointly by both Academies and the Institute of Medicine. Dr. Bruce M. Alberts and Dr. Wm. A. Wulf are chairman and vice chairman, respectively, of the National Research Council.

COMMITTEE ON THE FUTURE OF THE U.S. AEROSPACE INFRASTRUCTURE AND AEROSPACE ENGINEERING DISCIPLINES TO MEET THE NEEDS OF THE AIR FORCE AND THE DEPARTMENT OF DEFENSE

ROBERT R. EVERETT, *Chair,* MITRE Corporation (retired), Concord, Massachusetts
GORDON R. ENGLAND, *Vice Chair,* General Dynamics Corporation, Falls Church, Virginia (from September 1999 to May 2001)
KYLE T. ALFRIEND, Texas A&M University, College Station
OLIVER C. BOILEAU, JR., Consultant, Saratoga, Wyoming
MICHAEL P.C. CARNS, U.S. Air Force (retired), Center for Political Economy, Pebble Beach, California
RAYMOND S. COLLADAY, RC Space Enterprises, Inc., Golden, Colorado
JOHN W. DOUGLASS, Aerospace Industries Association of America, Inc., Washington, D.C.
ROBERT B. ORMSBY, JR., Lockheed Aeronautical Group Systems (retired), Roswell, Georgia
GEORGE A. PAULIKAS, Aerospace Corporation (retired), Los Angeles, California
THOMAS M. PERDUE, Consultant, Alexandria, Virginia
WINFRED M. PHILLIPS, University of Florida, Gainesville
HERMAN M. REININGA, Rockwell Collins, Cedar Rapids, Iowa
ELI RESHOTKO, Case Western Reserve University, Cleveland, Ohio
MICHAEL D. RICH, RAND, Santa Monica, California
HAROLD W. SORENSON, MITRE Corporation, Bedford, Massachusetts

Liaisons from the Air Force Science and Technology Board

ROBERT A. FUHRMAN, Lockheed Corporation (retired), Pebble Beach, California
JOHN MICHAEL LOH, U.S. Air Force (retired), Consultant, Williamsburg, Virginia

Liaison from the Board on Army Science and Technology

ALLEN C. WARD, Ward Synthesis, Inc., Ann Arbor, Michigan

Staff

BRUCE A. BRAUN, Director
JAMES E. KILLIAN, Study Director (from December 2000)
JAMES D. RENDLEMAN, Study Director (from September 1999 to October 2000)
LINDA D. VOSS, Consultant
JAMES MYSKA, Research Associate
PAMELA A. LEWIS, Senior Project Assistant (from January 2001)
ANDRE MORROW, Senior Project Assistant (from September 1999 to January 2001)

AIR FORCE SCIENCE AND TECHNOLOGY BOARD

Staff

Preface

The uncertainty of the threats faced by the military since the end of the Cold War is mirrored by uncertainties in the national defense aerospace infrastructure. The aerospace industry has undergone a significant restructuring in the last 20 years, a dramatic consolidation to adjust to the declining defense investment. In the 1980s, aerospace was a major U.S. economic sector dominated by defense spending. In the 1990s, the U.S. Department of Defense (DoD) accounted for only 28 percent of aerospace sales.

These changes raise questions about the future. The Air Force is concerned about having available and attracting the creative, skilled work force it will take to implement its military mission. The change in the environment supporting the defense aerospace infrastructure has also changed the relationship between the military and industry.

The committee was asked to identify problems facing different sectors of the defense infrastructure and how the Air Force could ensure its ability to attract the best and brightest to produce the leading-edge technology upon which its weapons systems rely.

To determine the scope of its study, the committee consulted with representatives of academia and visited representatives of the Air Force, Navy, DoD, Defense Advanced Research Projects Agency, National Aeronautics and Space Administration, Federal Aviation Administration, and other government agencies. To learn more about issues related to the physical test and development infrastructure, a fact-finding team visited the Arnold Engineering Development Center and received briefings about the work force, budget, policy, and facilities. Industry representatives made presentations on work force issues, business opportunities and goals, facilities, and financial challenges. In the end, the committee focused its attention on the issues most important for the primary client of the defense aerospace infrastructure, the Air Force.

Recommendations are focused on how Air Force senior management can compete for skilled technical personnel, sustain high-quality scientific and technical resources, and reform industrial policy to adapt to the changes in the industry.

The committee greatly appreciates the support and assistance of National Research Council staff members James Killian, Pamela Lewis, and Carol Arenberg and consultant Linda Voss in the production of this report.

Robert R. Everett, *Chair*
Committee on the Future of the U.S. Aerospace
Infrastructure and Aerospace Engineering
Disciplines to Meet the Needs of the Air Force
and the Department of Defense

Acknowledgments

This report has been reviewed in draft form by individuals chosen for their diverse perspectives and technical expertise, in accordance with procedures approved by the National Research Council's (NRC's) Report Review Committee. The purpose of this independent review is to provide candid and critical comments that will assist the institution in making its published report as sound as possible and to ensure that the report meets institutional standards for objectivity, evidence, and responsiveness to the study charge. The review comments and draft manuscript remain confidential to protect the integrity of the deliberative process. The committee wishes to thank the following individuals for their review of this report:

Henry P. Arnold, BF Goodrich Aerospace
William C. Bowes, USN (retired), Litton Integrated Systems
Natalie W. Crawford, RAND
Earl H. Dowell, Duke University
David Heebner, Heebner Associates
R. Richard Heppe, Lockheed Aeronautical Systems Company
Peter B. Teets, Lockheed Martin Corporation (retired)
Brian Wright, Rockwell Collins

Although the reviewers listed above have provided many constructive comments and suggestions, they were not asked to endorse the conclusions or recommendations, nor did they see the final draft of the report before its release. The review of this report was overseen by Alton Slay, appointed by the Division on Engineering and Physical Sciences, and William Howard, appointed by the NRC's Report Review Committee, who were responsible for making certain that an independent examination of this report was carried out in accordance with institutional procedures and that all review comments were carefully considered. Responsibility for the final content of this report rests entirely with the authoring committee and the institution.

Contents

Tables and Figures

TABLES

FIGURES

Acronyms and Abbreviations

ABL	airborne laser
AEDC	Arnold Engineering Development Center
AFB	Air Force Base
AFIT	Air Force Institute of Technology
AFMC	Air Force Materiel Command
AFRL	Air Force Research Laboratory
CAS	cost-accounting standards
CEO	chief executive officer
CFD	computational fluid dynamics
COTS	commercial off-the-shelf
DARPA	Defense Advanced Research Projects Agency
DASA	DaimlerChrysler Aerospace
DCS	deputy chief of staff
DoD	U.S. Department of Defense
DSB	Defense Science Board
EMD	engineering and manufacturing development
FAA	Federal Aviation Administration
FAR	Federal Acquisition Regulations
FFRDC	Federally Funded Research and Development Center
FSA	Future Strike Aircraft
FY	fiscal year
GOCO	government-owned, contractor-operated
GPS	global positioning system
IHPTET	Integrated High Performance Turbine Engine Technology
IR&D	independent research and development
ITAR	International Traffic in Arms Regulations
JPL	Jet Propulsion Laboratory
JSF	Joint Strike Fighter
LAI	Lean Aerospace Initiative
Mil-Spec	military specification
MOSA	modular open systems architecture
NAE	National Academy of Engineering
NASA	National Aeronautics and Space Administration
NATO	North Atlantic Treaty Organization
NMD	National Missile Defense
NRC	National Research Council
O&M	operations and maintenance
ODTC	Office of Defense Trade Controls
OPM	Office of Personnel Management
OSD	Office of the Secretary of Defense
QDR	Quadrennial Defense Review
R&D	research and development
RDT&E	research, development, test, and evaluation
RLV	reusable launch vehicle
ROI	return on investment
S&T	science and technology
SBL	Space-Based Laser
TINA	Truth in Negotiations Act
UAV	unmanned air vehicle
UCAV	unmanned combat air vehicle
USAF	U.S. Air Force

Executive Summary

BACKGROUND

In the 1980s, aerospace was a major U.S. economic sector dominated by defense spending. In the 1990s, the U.S. economy shifted gears, moving rapidly into new, nondefense sectors, such as information technology and biotechnology. In those 10 years, the aerospace industry was dramatically consolidated to adjust to a shrinking Department of Defense (DoD) market and decreasing defense investments. In 1989, DoD accounted for 51 percent of aerospace sales in the United States. In 1999, DoD accounted for only 30 percent of aerospace sales (AIA, 2000, 2001a). This change has altered the relationship between the military and the industry but has not been reflected in changes in policies and regulations relating to the defense industry.

The United States currently leads the world in aerospace technology. Academia continues to produce skilled aerospace engineers, and industry continues to produce excellent aerospace products. The question is how long this lead can be sustained in the new environment—how long DoD, and specifically the Air Force (which requested this study), can maintain its leadership in aerospace, continue to attract the highest-quality scientific and technical personnel for positions in both government and the defense sector of the aerospace industry, innovate faster than its potential adversaries, and maintain its military advantage.

ORIGINS OF THIS STUDY

The Principal Deputy to the Assistant Secretary of the Air Force for Acquisition requested that the National Research Council (NRC) review the Air Force's planned acquisition programs to determine if, given its scale, the highly talented scientific, technical, and engineering personnel base could be maintained, to identify issues affecting the engineering and science work force, and to identify issues affecting the aerospace industry's leadership in technology development, innovation, and product quality, as well as its ability to support Air Force missions.

A major concern of the Air Force is whether the aerospace industry can continue to produce cutting-edge products and attract the highly skilled technical people necessary for the industry to meet the Air Force's future needs. A related concern is the Air Force's ability to attract similar personnel to perform and manage research, technical, and acquisition programs as Air Force government service employees. Another concern is the economic health of the aerospace industry, particularly whether the level of funding for research and development (R&D) and science and technology (S&T) provided by the government and the commercial sector is adequate, whether the future of research and test facilities is in jeopardy, and whether U.S. superiority and leadership in aerospace can be maintained.

The Air Force's concerns are based on the following factors:

- Military budgets have been reduced substantially, resulting in fewer new programs and fewer career opportunities for new people.
- Industrial mergers have reduced the number of companies and potential suppliers; downsizing, with its attendant instabilities, has made working in the defense industry in general and the defense sector of the aerospace industry in particular (hereinafter referred to as the aerospace defense industry) less desirable.
- The independent research and development (IR&D) pool is now concentrated in fewer companies. Each remaining company concentrates its IR&D efforts on a single solution, thereby reducing the amount of total design effort.
- A large number of experienced technical people in both government and industry are approaching retirement age and must be replaced. Because of downsizing, however, relatively few new people have been hired in recent years. Thus, the depth of experience in middle management and the availability of qualified future senior managers are questionable.
- The growth of other commercial industry sectors,

spurred by rapid growth in information technology, has greatly increased the demand for technical people, especially software engineers.

STATEMENT OF TASK

The NRC was requested to:

- Gather information from key sources on the status of, and issues surrounding, the current aerospace infrastructure. This would include information gathering from government and national private sector stakeholders such as NASA, DARPA, FAA, warfighters, academia, senior industry executives, and military acquisition personnel.
- Examine component sources such as the aircraft engine industry that have been successful despite low and intermittent production rates to determine if lessons learned have wider applicability.
- Assess whether planned acquisition programs, considering their requisite S&T investment, will provide sufficient opportunities for innovation and to maintain a critical mass of activities to sustain a highly talented engineering talent base. Use planned aircraft and space vehicle development programs, major technology development and weapon system modification programs, and NASA programs as potential information sources. Consider international aerospace market sales prospects, as appropriate.
- Identify issues relative to the maintenance of an adequately educated, trained, and innovative force of engineering and science professionals to support the national aerospace infrastructure and on how the aerospace industry can maintain its world leadership in technology development, innovation, and product quality.

APPROACH TO THIS STUDY

To undertake this study, the NRC formed the Committee on the Future of the U.S. Aerospace Infrastructure and Aerospace Engineering Disciplines to Meet the Needs of the Air Force and the Department of Defense, under the auspices of the NRC's Air Force Science and Technology Board. The committee focused its attention on the trends and issues with the most serious impact on the defense aerospace sector of the aerospace industry, and the committee's recommendations are focused on remedies available to the Air Force. After reviewing the economic health of the aerospace industry; the Air Force and industry S&T talent base; the quality of test facilities and support capabilities; relevant DoD policies, regulations, and procedures; and management in principal government, industry, and academic organizations the committee made the following general observations:

- Although procurement funding levels have recently increased slightly, today's defense aerospace industry provides less work for aerospace engineers as a result of force drawdowns and reductions in procurement funds in the early 1990s. The military has fewer aerospace systems in development today than at any time in the recent past because of decreases in procurements; less apparent need for new systems since the end of the Cold War; and robust, life-extending techniques developed by industry for current aerospace systems that have given new life to older systems.
- The pool of scientific, technical, and engineering talent in the aerospace field is shrinking as a result of losses to highly attractive, competitive industries. The military and the aerospace defense industry face similar recruitment challenges.
- The Air Force does not have sufficient resources to support all of its missions and programs. This has caused excessive funding uncertainties and program instabilities.
- Many commercial firms, both established and new, are reluctant to enter into government contracts (other than commercial sales) because of perceived "unreasonable" government practices.
- The government service personnel system, which has rigid salary structures and complex, time-consuming hiring practices, has made meeting the needs of government research laboratories extremely difficult.
- Because a unique characteristic of the aircraft engine industry is that defense products can piggyback onto its commercial products, this sector is not a good model for the aerospace industry as a whole.

THE BASIC CONCLUSION

The technical resources problem is not separable from the Air Force's other duties. It affects and is affected by Air Force decisions about current and future missions and needs and what the Air Force develops, buys, tests, and uses in training, which in turn leads to what the future of the Air Force is to be. If the technical resources required are not considered when and where these fundamental matters are discussed and decided, their effects will not be properly taken into consideration. The Air Force pays close attention to those matters it holds most important. For many years, however, it enjoyed high-quality technical resources without paying specific attention to them, but times are changing. In the opinion of the committee, the technical resources will not continue to be of high quality without this specific attention.

In the committee's opinion, this problem is best dealt with by the Chief of Staff of the Air Force saying he wants attention paid to technical quality and quantity and then appointing a deputy for this purpose, preferably as a sole responsibility.

Further, although S&T is important, in fact very important, it is only a part of what concerns the committee. The committee believes it would be desirable for a deputy chief of staff (DCS) to have responsibility for S&T as a part of his larger job. Alternatively, someone else could have direct S&T responsibility, as is true for most of the other elements of the DCS's portfolio. This DCS, in the committee's view, would have oversight of all Air Force technical resources and how they fit into what the Air Force is doing and plans to do.

The basic conclusion of this report is that if the Air Force is concerned about the future of its technical resources (and it is), it must give the problem the kind of continuing attention that it gives to other serious matters. This attention includes the following:

- Raising the level of attention by establishing a deputy chief of staff who is also a member of the Air Force Council to oversee the Air Force's technical resources;
- Creating an ongoing assessment of Air Force technical resources and planning what to do to ensure their quality;
- Paying attention to all elements of the technical spectrum, including S&T, which is necessary although not sufficient, as well as R&D, design, and production;
- Considering separately the health of industry and inhouse labs, the efficacy of program management, and the health of universities—and doing what is necessary to ensure their continued health in supporting the Air Force; and
- Making sure the technical community fully understands the Air Force's need for and commitment to high-quality, leading-edge technology and the role of Air Force technology and acquisition people.

SPECIFIC CONCLUSIONS AND RECOMMENDATIONS

The character of potential adversaries has changed, as has the environment of the defense aerospace industry. For the Air Force to maintain its technological lead, it too must change if it is to maintain the quality of its technical resources. The major areas of change should be in the management of scientific and technical resources, maintenance of the technical work force, the Air Force's relationship with industry and academia, and related government industrial policies. The committee's major conclusions and recommendations in these areas are presented below.

Scientific and Technical Resources

A shrinking budget and some insistence on maintaining business as usual have led to great instabilities. The Air Force has more programs, particularly new procurement programs, than it can support with its present and foreseeable acquisition budgets. As a result, funds are constantly being shifted and reprogrammed to pay pressing bills. This has created unnecessary, debilitating upheavals for both personnel and programs and may prejudice the Air Force's long-term technology initiatives.

The committee recognizes that the Air Force major program budget is heavily affected by the political process. Nevertheless, the Air Force should decide upon, and then protect, the portion of the budget allocated for future technologies, which will determine the quality of its future warfighting capability.

Conclusion 1. In the process of refocusing its priorities and as a result of reorganizations predicated on the Goldwater-Nichols Act, the Air Force eliminated the position of Deputy Chief of Staff for Research and Development and thus lost a strong advocate for science and technology. In addition, the Air Force Systems Command was combined with the Air Force Logistics Command to form the Air Force Materiel Command (AFMC). Although this consolidation has streamlined AFMC's processes for development, acquisition, and support of Air Force systems, it has also reduced the emphasis on technology in general and S&T in particular. Currently, the highest S&T-dedicated position in the Air Force is the two-star Air Force Research Laboratory (AFRL) commander position at Wright-Patterson Air Force Base (AFB) near Dayton, Ohio, which is several levels below the Air Force Council. The AFRL commander reports directly to a general (four-star), the commander of AFMC, of which AFRL is a part. AFMC headquarters is also located at Wright-Patterson. The AFMC commander's responsibilities are very broad, including the programs at four product centers, five air logistics centers, three test centers, and two major specialized centers, in addition to AFRL. The AFMC commander has too many other important and demanding responsibilities to focus on S&T, and without an S&T advocate of sufficient stature and authority at the Air Staff level where budget decisions are made, support for S&T has declined substantially. Reinstating a senior voice for technology in shaping the future capability of the Air Force would help ensure that decisions affecting technical resources including S&T expenditures are fully considered by Air Force decision makers.

Goldwater-Nichols defines the relationship between the Secretary's Office and the Chief's Office on technical matters. The Secretary's responsibilities are clear, but this should not mean the military does not have very strong concerns about and influence on the technical resources of the Air Force and does not have the need for oversight. The committee believes that the Air Force, both civilian and military, must pay more focused attention to its technical resources if it is to continue to get the best weaponry.

If Goldwater-Nichols or other constraints make a DCS position unworkable, the role and responsibility recommended could be assigned in other ways. The committee understands, for example, that the Navy has established a position for a two-star Director of Test & Evaluation and Technology Requirements (N091) who reports directly to the Chief of Naval Operations (CNO) and has somewhat similar functions. He serves as the principal interface between the CNO and the Assistant Secretary of the Navy for Research, Development and Acquisition on RDT&E; Resource Sponsor for Navy S&T (6.1, 6.2, 6.3A) investments; Resource Sponsor for RDT&E field activities; and Appropriations Sponsor for CNO RDT&E, Navy (RDT&E,N) funding. In any event, the committee believes that high-level attention is needed to get the best result.

Recommendation 1. The Air Force should establish a deputy chief of staff, who is also a member of the Air Force Council, with primary responsibility for oversight of all Air Force scientific and technical resources. Among his duties, which should include all Air Force technical activities from concept development through completion of engineering and manufacturing development (EMD) phases, this officer should be the advocate for funding science and technology requirements and for modifying and tracking the implementation of S&T requirements to minimize instabilities in S&T and R&D funding (including new production processes), to ensure that adequate funding is budgeted annually, and to resist attempts to raid S&T or R&D funds to meet short-term budget shortfalls in other areas.

The committee believes that whoever is assigned the responsibilities for oversight of Air Force technical resources should be able to act as a high-level advocate for the technical resources within the Air Staff and the department. This individual should be someone who has an extensive scientific or technical education and background, yet also has experience in the operational commands and can appreciate the critical needs of both sides of the house—warfighting and technical. The new DCS would maintain awareness of the status of all aspects of the Air Force's technical resources and would track the effects of current and proposed policies—concerning personnel, facilities, the Air Force Institute of Technology, education, and research activities—on the technical capabilities base. This person would then serve as an advocate on the Air Staff to ensure that the needs of the technical capabilities base at least get a fair and accurate hearing in the policy decisions of the Air Force.

The Air Force Technical Work Force

Despite the recent difficulties of attracting qualified people to government defense work, the committee believes the Air Force has marginally enough scientific, technical, and engineering personnel to carry out its current programs and, with the appropriate effort, should be able to attract enough people in the near future to develop and build the systems that are now planned. In fact, the Air Force, even with reduced S&T funding, has the resources to pursue many important programs on the leading edge of technology, providing it sets the right priorities and executes them efficiently. The problem is how to attract new talent in the face of growing commercial competition. This is a new challenge for the Air Force, and meeting it will require a new way of thinking. The committee strongly supports technical education for Air Force personnel at both the Air Force Institute of Technology (AFIT) and civilian universities as a source of technically educated officers for S&T and acquisition force positions.

Conclusion 2. Technical personnel—having the types and numbers that are essential for maintaining current and future Air Force technological superiority and for ensuring technical innovations in science, technology, research, and development to support current, emerging, and future capabilities—are just as important to the successful accomplishment of the Air Force mission as the operational and support elements.

Recommendation 2. The Air Force should assess the quality and quantity of its technical personnel regularly, taking into consideration its future missions and needs, just as it currently assesses the quality and quantity of its flight personnel and other vital resources. The Air Force should use these assessments to define the types and numbers of technical personnel necessary to maintain current and future Air Force technological superiority and ensure technical innovations in science, technology, research, and development. The results of the assessment should be used as a basis for making policy changes (if necessary) to protect investments that support technical personnel.

Ongoing assessments should include the following estimates:

- The number of uniformed and civil service technical personnel necessary now and in the future, including skills, technical specialties, and years of experience;
- The organizational base and operating requirements to support the Air Force's long-term technology needs, including size, skills, and responsibilities for universities and in-house laboratories performing scientific and technical research, industrial contractors that the Air Force considers essential to maintaining a competitive supplier base, organizations the Air Force intends to sustain as sole suppliers in particular areas, and Federally Funded Research and Development Centers (FFRDCs); and
- The need for research, development, test, and evaluation (RDT&E) facilities, such as wind tunnel facilities and test ranges, for transitioning technology capabilities and meeting research requirements. This is included because high-quality technical people need and will insist on high-quality facilities. These needs should be closely coordinated with the other services and with the National Aeronautics and Space Administration (NASA).

These assessments should be the responsibility of the deputy chief of staff recommended above. Finally, that officer should present the assessments of Air Force technical resources, both current and future, to the senior Air Force leadership annually during the periodic meeting at Corona.

Conclusion 3. With a nearly 35 percent drop in the Air Force S&T budget over the last 15 years, the Air Force cannot sustain its technical viability over the long term. Although total S&T funding for DoD as a whole is on the order of 4 percent of the DoD budget, the Air Force has selectively reduced the percentage it allocates for this purpose compared to the other services. The Air Force reductions in S&T funding also come at a time when other sources of federal S&T funding that support the Air Force, such as NASA and Federal Aviation Administration research, have declined significantly. S&T funding is the "seed corn" for the technical superiority of the Air Force. Therefore, the level of S&T funding must be maintained at an adequate level.

Recommendation 3. The Air Force should balance current expenditures and investments in future technologies and insulate those budgets from the vagaries of near-term fiscal pressures. Vital science and technology resources should be organized, protected, and nurtured just as carefully as critical operational resources.

Relationship with Industry

Few, if any, perfectly free markets exist anywhere with many suppliers and many buyers, perfect information, and no applied restraints. The DoD as a monopsony, or single buyer, for the defense industry cannot be said to operate in anything like a free market. This, however, does not mean that there is no competition, just that the competitions are established and controlled by DoD. The DoD has widely varying relationships with its suppliers, ranging from open competitions to what are essentially permanent single sources and all other combinations in between. Since DoD sets the rules, it is responsible for the effects of those rules on its supplier base whether it recognizes this explicitly or not. The committee understands that this is well recognized within DoD but that there seems to be no established mechanism for determining and taking into account the effects on the Air Force's technical resources when decisions are being made about the competitive conditions for individual programs or for the Air Force program as a whole.

In a free market, the responsibility for maintaining a high-quality technical staff can and should be left to suppliers. However, the aerospace defense industry is not a free market. The Air Force and DoD as a whole are a monopsony customer for much of the aerospace industry and therefore have significant influence over the supplier base and infrastructure. Despite changes in DoD policy requiring that all of the services increase their use of commercial off-the-shelf products and modify their procedures to accommodate commercial practices, the Air Force still depends on the aerospace defense industry for high-quality, Air Force-unique products. Through its procurement practices, the Air Force can affect, even determine, the long-term viability of these suppliers.

Conclusion 4. In the current environment, the Air Force cannot continue to think of the defense aerospace industry as a competitive marketplace. The way the Air Force uses its resources, by design or otherwise, has a major impact on the viability of the defense aerospace infrastructure that supports the national security interests of the Air Force. Maintaining the traditional arm's-length relationship with industry must yield to establishing long-term partnerships with responsibilities on both sides. Partnership includes the concepts of mutually supportive and collaborative relationships and two-way exchange in which the parties depend on each other. Industrial organizations must be responsible for maintaining their own capabilities, but the Air Force must be responsible for providing conditions and incentives under which these organizations can remain strong and effective and continue to enhance their technical capabilities.

Recommendation 4. Air Force management should take into consideration the effects of its budget and management policies on industry. The Air Force should establish partnerships with defense industries that will encourage industry's continued effectiveness and enhance its technical capabilities.

A number of studies analyzing the defense industrial environment were reviewed by the committee. These studies largely agree that the needs of the industrial base should be taken into account in the acquisition process, that industry metrics must be better understood, and that the export control process must be streamlined. The committee's investigations substantiated the results of these studies and the validity of their recommendations.

Although government oversight of the defense industry is necessary to ensure that taxpayer money is spent wisely and appropriately, current government regulations are focused excessively on detailed cost accounting and adherence to complex processes. As a result of these regulations, efficiency has been undermined, procurement costs have risen substantially, and the attractiveness to technical people of working in defense has been reduced. Unless high-quality people are treated as trustworthy and reliable, they will go elsewhere. Some leading technology corporations have divested themselves of divisions and subsidiaries that made defense products. Although most of these divisions and subsidiaries were acquired by companies that made similar or related products, the number of companies bidding on Air Force contracts has been reduced.

The Air Force would benefit a great deal if it could take advantage of commercial technology and processes. New initiatives for modifying the Federal Acquisition Regulations (FAR) should enable a shift toward more commercial-like procurement relationships.

Conclusion 5. Current government reforms are a step in the right direction toward dramatically reducing acquisition cycle times and will certainly improve the efficiency of the defense industry. They will also allow greater flexibility in program design and result in processes that are more in tune with industry. Reform should be considered a long-term undertaking and should continue to be visibly supported by Air Force leadership. The issue of whether reducing cycle time will increase near-term budgets depends on whether one is attempting to squeeze all of the existing programs into a shorter time or whether one is trying to do as many as can reasonably be done within the available funds. Programs that cannot be fitted in will be delayed. Over a long period, more programs would get done within the same total funds, most of them sooner than they could be now because of the shortened cycle time and greater efficiency.

Recommendation 5. The Air Force, and the U.S. Department of Defense as a whole, should continue to provide strong leadership for initiatives to reduce acquisition cycle times as a means for furthering DoD goals. They should also continue to work toward reforming policies and regulations for acquisition processes, thereby reducing the burden on industry of working with the government.

Reforming Policy and Regulations

In addition to interesting work, quality people need quality working conditions. A healthy organization, whether in industry or in government, must provide competitive salaries, benefits, and opportunities for growth and advancement. Civil service regulations, which apply to the government service white-collar administrative force, the blue-collar work force, and the technical work force, use standardized processes that are poorly suited to meet the demands of government laboratories performing leading-edge research. Inflexible civil service regulations, hiring practices, employment conditions, and salary structures have resulted in a serious decline in government technical talent, and more and more bench-scale technical work is being contracted out to industry and university laboratories.

This adverse trend has exacerbated the problem of the viability and sustainability of government laboratories. Leaders of government technical organizations have been working hard to overcome these difficulties but have not received adequate policy and regulatory support. In the committee's opinion, the supply of good technical people the Air Force needs cannot be maintained under existing civil service rules and restrictions. Therefore, the Air Force must actively pursue civil service reform for scientific and technical personnel. A Defense Science Board (DSB) Task Force addressing the ability of DoD to attract and retain critical personnel recommended the transfer of authority for the DoD civilian work force from the Office of Personnel Management (OPM) to the Secretary of Defense (DSB, 2000a). The committee did not endeavor to evaluate this proposal but believes that this concept merits serious consideration.

Conclusion 6. The Air Force is facing serious competition for high-quality technical people. Although the Air Force can offer interesting and important work for in-house government employees and has a substantial budget, it has had trouble attracting people either directly out of universities or from the commercial sector. Many of the difficulties are related to civil service rules and restrictions that do not allow the salary flexibility and rapid decisions necessary to hire talented available applicants quickly. In order to meet the need to compete more effectively than it does now, the Air Force must provide a quality workplace.

Recommendation 6. The Air Force should join the Office of the Secretary of Defense, the other services, and other federal agencies in ongoing attempts to reform the civil service rules for scientific and technical personnel.

Universities are the trainers and motivators, the breeding ground, of the future defense work force. Air Force funding for university research is essential for maintaining this connection. The Air Force should cultivate and establish long-term partnerships with universities.

Conclusion 7. The universities, faculties, and students whose research is supported by U.S. government funds are an indispensable base for motivating young, well-educated individuals to pursue careers in defense technology and Air Force laboratories. The relationship between universities and the Air Force is symbiotic. Universities need S&T funds from the Air Force, and the Air Force needs both the results of S&T and new technical personnel. If the Air Force wants a healthy relationship it must cultivate and establish long-term partnerships with universities.

Recommendation 7. The Air Force should establish long-term, stable partnerships with its supporting universities and their faculty members. The Air Force should decide how much to invest for the future through S&T funds to universities and then protect that investment. The Air Force should also recognize the financial problems facing universities and make sure that contractual and financial arrangements are consistent with the continued health of these important institutions and their ability and willingness to continue to support the Air Force.

The Air Force must do more to attract and retain the highest-quality scientific and technical people. As a matter of first principle, highly trained technical people want interesting and important work, which the Air Force has in plentiful supply. Despite reduced R&D budgets, the Air Force continues to devote major funds to solving leading-edge technical challenges. However, the impression in industry and the technical community is that the Air Force is short of money and is concentrating its efforts on operational requirements, modifications, and upgrades at the expense of new technology. Although the Air Force's emphasis has recently been focused on operational requirements for multiple force deployments, substantial funds are still being invested in interesting and challenging S&T projects. The Air Force should develop a strong, positive message about its technical S&T program.

Conclusion 8. Air Force technical programs and opportunities are challenging and exciting. However, the Air Force has not communicated that excitement to the technical community. The Air Force must overcome the perception that opportunities in defense research are limited and that defense is not as important as it was during the Cold War.

Recommendation 8. The Air Force should communicate a strong, positive message describing its technical plans and opportunities and ensure that this message is broadly distributed to students, faculties, industry, and the general technical community.

1

Introduction

BACKGROUND

For many years the aerospace infrastructure has been supporting the development and production of aerospace products critical to our national security and economy. U.S. industry has dominated aviation and aerospace for most of its history. Commercial aerospace exports, which are dependent in large measure on the technology base that is in turn supported by the Air Force, are traditionally by far the largest positive contributor to the U.S. balance of trade (Figure 1-1).

Cutting-edge aerospace products will continue to be essential to U.S. dominance of the twenty-first century battlespace. In response to many stimuli—the end of the Cold War, reduced military budgets, increased commercial competition, and a shrinking market—the aerospace industrial base has undergone extensive consolidation. The ramifications of this consolidation for the military include uncertainty about maintaining sources of supply and sustaining a base of world-class engineers to design and produce future military systems. Although the U.S. market share of weapons exports has increased recently, there is no guarantee that this trend will continue. In fact, the overall positive aerospace trade balance has fallen by about 35 percent since 1998 (Douglass, 2000b).

The Air Force has also expressed concerns about whether the Air Force and the aerospace industry can continue to attract and retain the numbers of highly skilled technical people they believe meeting their future needs demands, both in government and in commercial industry. These concerns are based on several factors:

- Military budgets are down substantially, resulting in fewer new programs and less resources to invest in science and technology (S&T).
- A large number of experienced technical people in both government and industry are close to retirement age and must be replaced. Because of downsizing, relatively few new people have been hired in recent years, raising concerns about the depth of experience in middle management and the availability of replacement senior managers.
- The strength of other industrial sectors, especially the emergence of the fast-growing information industry, has greatly increased the demand for technical people, especially software engineers.

The Air Force and the other services are assessing the changes in the defense aerospace infrastructure and the effects on development and support of future military systems. The Air Force is looking for ways to interact with and provide incentives to industry to ensure that national needs will be met well into the twenty-first century.

STATEMENT OF TASK

In discussions with the Air Force, the National Research Council (NRC) was asked to provide a report that addresses the effects of U.S. defense industrial base shrinkage and the aerospace industry's ability to continue to attract and maintain requisite aerospace engineering talent to be able to produce cutting-edge military products to support the Defense Department's (DoD's) needs. The report would make recommendations to DoD, and in particular the Air Force, on ways to keep the defense industry, and its aerospace engineers, on the forefront of technology and to maintain its capacity to innovate through enhanced practices, policies, and procedures.

The NRC was asked by the Principal Deputy to the Assistant Secretary of the Air Force for Acquisition to assist in the assessment of the defense aerospace infrastructure and identify future trends. In response, the NRC formed the committee on the Future of the U.S. Aerospace Infrastructure and Aerospace Engineering Disciplines to Meet the Needs of the Air Force and the Department of Defense, under the auspices of the Division on Engineering and Physical Sciences. The NRC was requested to:

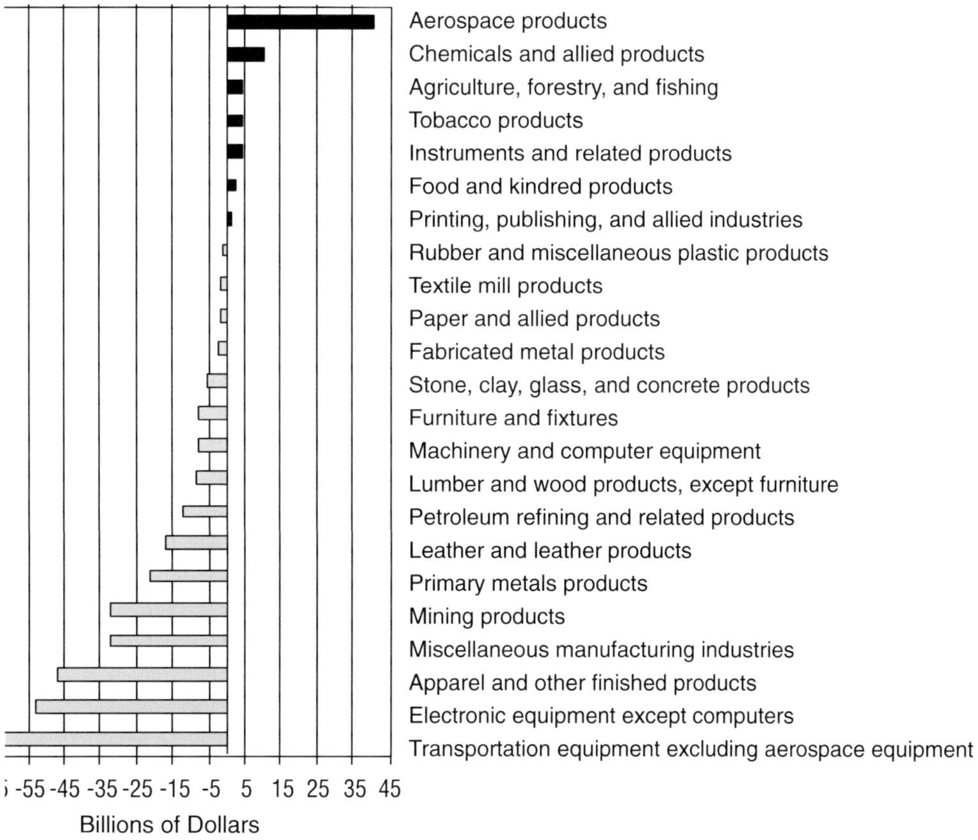

FIGURE 1-1 Balance of trade by industry, 1998. SOURCE: Douglass, 2000a.

- Gather information from key sources on the status of, and issues surrounding, the current aerospace infrastructure. This would include information gathering from government and national private sector stakeholders such as NASA, DARPA, FAA, warfighters, academia, senior industry executives, and military acquisition personnel.
- Examine component sources such as the aircraft engine industry that have been successful despite low and intermittent production rates to determine if lessons learned have wider applicability.
- Assess whether planned acquisition programs, considering their requisite S&T investment, will provide sufficient opportunities for innovation and to maintain a critical mass of activities to sustain a highly talented engineering talent base. Use planned aircraft and space vehicle development programs, major technology development and weapon system modification programs, and NASA programs as potential information sources. Consider international aerospace market sales prospects, as appropriate.
- Identify issues relative to the maintenance of an adequately educated, trained, and innovative force of engineering and science professionals to support the national aerospace infrastructure and on how the aerospace industry can maintain its world leadership in technology development, innovation, and product quality.

STUDY APPROACH

The committee collected information from a multitude of sources to identify issues and assess the status of the aerospace infrastructure. The committee examined the financial health of the industry across its full spectrum from research and development (R&D) through production and operations and maintenance (O&M); S&T talent base; key facilities and support capabilities; associated policies, regulations, and procedures; and management of the relevant government organizations and representative segments of industry and academia. The committee focused on those issues with the greatest effect on the defense aerospace sector of the aerospace infrastructure, which includes government laboratories and facilities, industry, and academia.

Committee members included experts and industry leaders with substantial experience in research and technology development programs and the production of aerospace systems, including academic R&D; industry management and manufacturing; aerospace, systems, and general engineering; avionics design and engineering; computer science; military procurement and contracting; control systems; information systems; and military operations, capability needs, and plans.

Over the course of nine meetings (see Appendix A) and many fact-finding site visits, the committee gathered information from representatives of the Air Force, Navy, DARPA, NASA, FAA, and other government agencies. Briefings by representatives of industry, academia, and government research laboratories described existing programs and facilities, the present and planned work force, military needs, and plans. To learn more about issues related to physical test and development infrastructure, a fact-finding team visited the Arnold Engineering Development Center (AEDC), where the team was given briefings about the work force, budget, policy, and facilities. The committee also obtained information from industry representatives on the work force, business opportunities and goals, facilities, and financial challenges. AEDC is a government-owned, contractor-operated (GOCO) facility; however, the committee did not pursue any "what-if" drills over the GOCO issue.

THE DEFENSE AEROSPACE INFRASTRUCTURE AND NATIONAL SECURITY

In the course of the study, the committee developed a framework for assessing the defense aerospace industrial base, especially the science and engineering work force. Vigorous technical progress in aerospace systems will be essential for implementing the national security strategy. The defense aerospace industrial base must be robust to produce the necessary technical progress. Even though the current technical capability of U.S. aerospace forces, demonstrated in Iraq and Kosovo, is very high, it will not remain so without continued investment in new technologies.

A robust defense aerospace industrial base requires adequate funding for research, development, test, and evaluation (RDT&E); a full spectrum of activities, from S&T and R&D through production and O&M; a highly trained, highly motivated technical work force; and effective policies for allocating and administering funding.

In the post-Cold War world, regional and ethnic rivalries and national and transnational terrorist and criminal organizations can present new, unexpected challenges to U.S. forces. The enemy in these situations may not be able to match U.S. military technology but will have two advantages. First, the enemy will have access to new commercial technologies with potential military applications, especially technologies in computers and communications, biotechnology, miniaturized electronics and surveillance equipment, and commercially available remote sensing. Casualties inflicted by this capability on U.S. forces can undermine public support far out of proportion to their military impact. Growing expectations that even relatively few casualties are unacceptable have created unique vulnerabilities and increased pressures on U.S. security policy. Second, an enemy can develop tactics specifically to counter U.S. advantages. These "asymmetric strategies" could include information net attacks, jamming and spoofing of the global positioning system (GPS), causing deliberate civilian damage to themselves to win international sympathy, and so on.

The unpredictable identity of the enemy, the access to rapidly changing commercial and military technologies, and the continual development of new tactics have created a new national security imperative: the United States must continue to improve its own technical capabilities to ensure that it can deter, and if necessary prevail, regardless of who the enemy is and what technologies or tactics are used. The committee believes that to maintain the agility of U.S. defense capabilities, and to stay ahead in the tactic and countertactic evolution, the United States must innovate faster and more effectively than potential enemies. Current DoD plans should guarantee this, but in the opinion of the committee they do not. DoD must have a deliberate policy, with adequate resources, to support and sustain a world-class defense aerospace industrial base.

Forces in competition with the United States, both friendly and unfriendly, will continue to invest in new technology. If we do not do the same, in time their weapons will be comparable, and then superior, to U.S. weapons. The United States would first lose its markets and then wars. DoD policy must be take into account the health of the industrial base that produces new technologies. At the current budget level of almost $300 billion (projected to remain about constant or possibly increase modestly in real dollars), technical progress will continue to be made, but not necessarily cutting-edge technical progress. To improve that probability, DoD must allocate and manage its finite resources more effectively. Inefficient management will mean less technical progress for the dollar. Equally important is the effect of inefficient management on the quality of the defense aerospace technical work force, in terms of both experience levels and the attractiveness of working in defense aerospace.

During the Cold War, the main focus of DoD funding was on improving the performance of weapon systems. Today, affordability is just as important, for both existing and emerging systems. All weapons, including aircraft, wear out and must be replaced and upgraded as technology advances and opportunities to improve systems performance arise. Rising costs can be offset only partially by the increased use of commercial technology. Air Force programs, in particular, involve some of the most sophisticated new technologies. To get the most out of its defense dollars, the Air Force must ensure that scientists and engineers are highly skilled, well educated, highly motivated, and experienced.

ORGANIZATION OF THIS REPORT

Chapter 2 discusses the issue of the level of S&T and R&D devoted to advancing technology. Chapter 3 contains a discussion of the quality and motivation of the defense aerospace work force and an assessment of planned aerospace modification, development, and acquisition programs in terms of S&T investments and whether they will support an innovative work force. Chapter 4 addresses the financial health of the aerospace industry. Chapter 5 covers the effects of policies on the allocation and management of resources, DoD's relationship with defense industries, the export of military items, and issues related to test facilities. Chapter 6 summarizes the committee's conclusions and recommendations.

2

Science and Technology Budgets

A robust defense aerospace industrial base requires a healthy science and technology (S&T) and research, development, test, and evaluation (RDT&E) funding stream. The future of the Department of Defense (DoD) and Air Force aerospace infrastructure will depend on investments in S&T and research and development (R&D).

The RDT&E budget comprises seven budget activities: (1) basic research, (2) applied research, (3) advanced technology development, (4) demonstration and validation, (5) engineering and manufacturing development, (6) RDT&E management support, and (7) operational systems development. These are also referred to as 6.1 through 6.7, respectively. The first three of these budget activities—basic research, applied research, and advanced technology development (6.1 through 6.3)—constitute the S&T budget. DoD's S&T budget is currently about 20 percent of the larger RDT&E budget, which includes the labor- and material-intensive tasks of incorporating new technical developments into military systems such as aircraft, avionics, engines, spacecraft, missiles, and weapons. The RDT&E budget is about 14 percent of the total DoD budget.

Industry S&T activities that can be charged as independent research and development (IR&D) are included in the RDT&E and overall procurement budgets. The total DoD S&T appropriation for FY01 was $9.0 billion, $1.5 billion higher than the $7.5 billion budget request submitted to the Congress. This S&T budget represents 3 percent of the overall DoD budget, and IR&D accounts for about another 1 to 2 percent. The total S&T budget is therefore about 4 percent of the DoD budget. This amount has a very great leverage as it has led in the past to the technical superiority of U.S. forces (DDR&E, 2001).

Historical trends in total DoD S&T and Air Force S&T funding are shown in Figures 2-1 and 2-2, respectively. In constant year FY01 dollars, the total DoD S&T budget rose from $6.6 billion in FY80 to a peak of $9.9 billion in FY93 (DDR&E, 2001). Since then it has declined somewhat and increased to $9 billion in FY01.

The Air Force S&T budget was approximately $1.8 billion in FY80, rose to a peak of $2.3 billion in FY88, and since that time has fallen dramatically, before rising slightly to $1.46 billion in FY01 (less than 17 percent of the total DoD S&T FY01 budget). This FY01 budget is only $1.42 billion if you do not count the 6.4-phase Space-Based Laser (SBL) funds moved into the S&T budget line in FY 2000 (USD(C), 2000a; DDR&E, 2001).

The total DoD budget, which peaked in 1985 at $436 billion in constant year FY01 dollars, has since decreased by nearly 33 percent, to $296 billion in FY01 (USD(C), 2000b; Heeter and Kosiak, 2000). During the same period, procurement spending has fallen by just over 56 percent, to $60 billion, and RDT&E spending fell by about 10 percent, to $41 billion (USD(C), 2000b; Public Law 106-259, the Department of Defense Appropriations Act, 2001). As the data show, the decreases for R&D were not nearly as large as those for other categories over this time. During this period, total DoD S&T actually increased, rising from $6.4 billion to $9 billion. In sharp contrast, the $1.5 billion of FY01 Air Force S&T represents a 25 percent decrease since FY85 and a 35 percent decrease since its peak in FY88 (DDR&E, 2001). Over the same period, NASA investment in S&T also declined (NRC, 1999). NASA's S&T is increasingly being focused on meeting the needs of specific NASA space programs and pulling back from its historic support for the development of aeronautics technologies. Also, NASA and the Air Force have not coordinated these budget reductions in aeronautical S&T investments and programs (Venneri, 2000).

The defense aerospace industry, too, has reduced its R&D investment. In response to the financial instability of the industry, businesses have focused on increasing shareholder value. Company R&D funds have been focused on near-term market prospects (analogous to 6.3 and 6.4). As a result, 6.1 and 6.2 projects have largely become the domain of universities and some military laboratories.

As a result of the demise of the Soviet Union and the

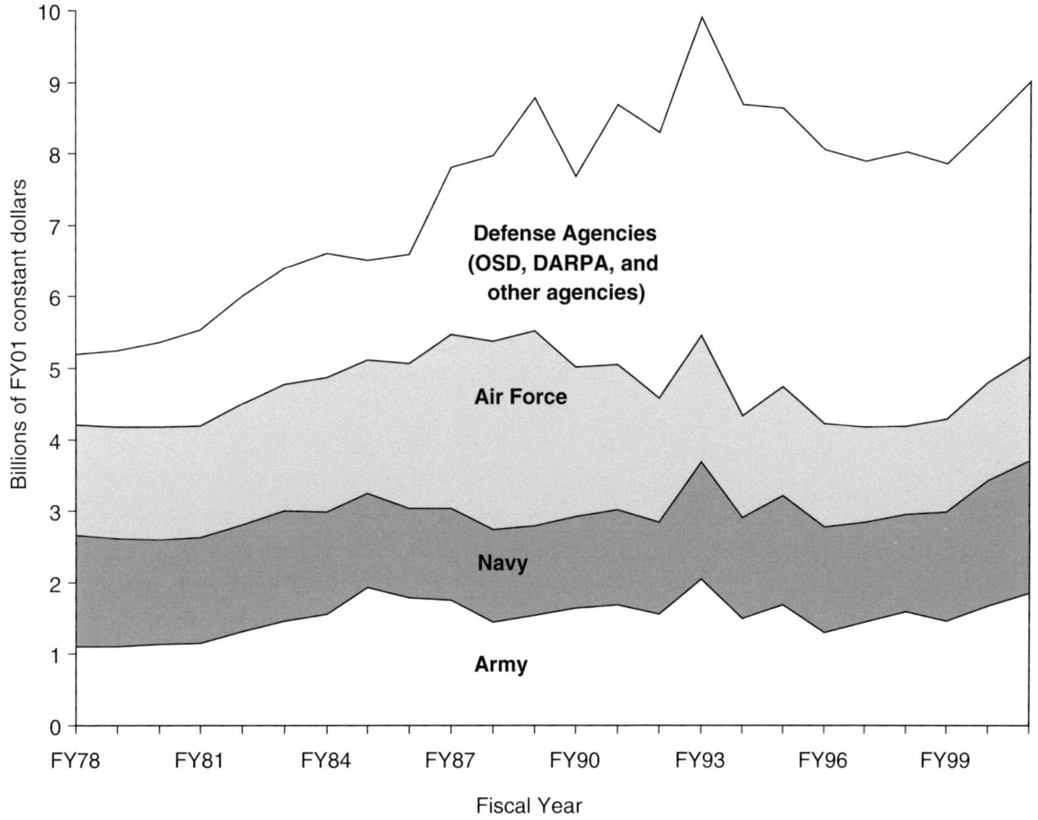

FIGURE 2-1 Total DoD S&T budget history. SOURCE: DDR&E, 2001.

subsequent downsizing of the U.S. military, the services have attempted to meet the resultant resource shortfall by taking advantage of excess inventory to address their equipment needs from existing systems or by modernizing existing systems rather than procuring new systems. In this light, the overall pattern of DoD budget reductions seems reasonable. However, RDT&E budgets, particularly the S&T portion of these budgets, must continue to anticipate the need for new technologies and systems once excess inventory has been depleted. Also, the change in threat places new demands on RDT&E.

The 35 percent decline in the Air Force S&T budget since 1988 is inconsistent with the Air Force goal of maintaining technological superiority. S&T funding must be maintained if the Air Force is to retain its technical superiority. A vigorous S&T program is also a necessary, but not sufficient, condition for the future health of the defense aerospace industry. Decisions about current expenditures and investments in future technologies should be based on a vision and a plan to ensure the development of technical resources for the future and to insulate S&T budgets from the vagaries of market fluctuations and short-term fixes.

INSTABILITY OF PROGRAM BUDGETS

The negative effects of budget reductions have been magnified by the instability of individual program budgets. Instability results from a policy of seeking to keep all or nearly all programs alive while shuffling money among them to deal with immediate problems. Shifting priorities and inadequate management reserves also disrupt programs, undermining the effectiveness of S&T and R&D spending.

Also, erosion in the integrity of the 6.1 to 6.5 funding process has destabilized S&T programs. Air Force S&T accounts have been reprogrammed to pay bills and to meet pressing O&M and procurement needs that have arisen with the recent increase in operations tempo (Etter, 2000).

Funding of Air Force S&T programs has become increasingly unstable. "Raiding" of S&T budgets to meet short-term readiness goals is an understandable reaction to short-term pressures but is not an acceptable policy for meeting long-term national security requirements because it destroys S&T program integrity and viability, wastes resources, and undermines the stability of the defense aerospace industry. Senior Air Force leadership must take into account that a viable industry is an absolute requirement for a strong U.S.

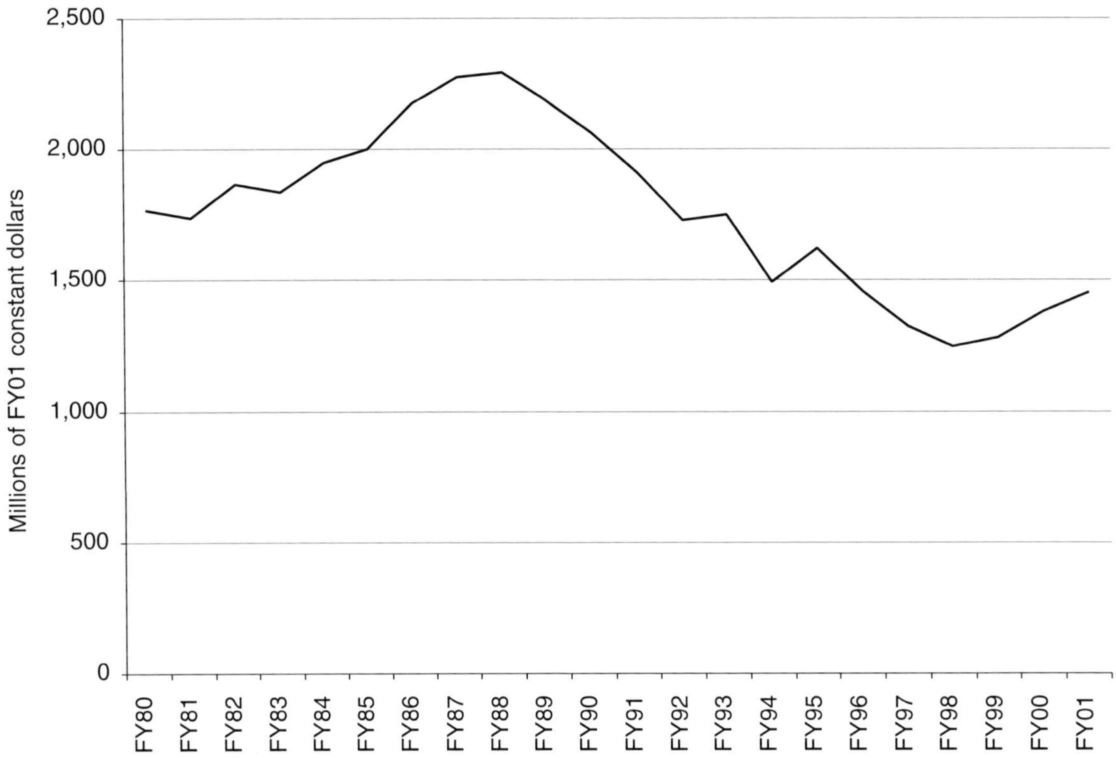

FIGURE 2-2 Air Force S&T budget history. SOURCE: DDR&E, 2001.

defense and that S&T funding is the source of technology advancement in the industrial sector.

In the process of refocusing its priorities and as a result of reorganizations predicated by the Goldwater-Nichols Act, the Air Force eliminated the position of deputy chief of staff for research and development, which served as a strong advocate for science and technology. In addition, Air Force Systems Command has been combined with the Air Force Logistics Command to form the Air Force Materiel Command (AFMC). Although this has streamlined AFMC's processes for development, acquisition, and support of Air Force systems, it has also tended to reduce the emphasis on S&T. Currently, the highest S&T-dedicated position in the Air Force is the two-star Air Force Research Laboratory (AFRL) commander position at Wright-Patterson Air Force Base (AFB) near Dayton, Ohio, which is several levels below the Air Force Council. The AFRL commander reports directly to a general (four-star), the commander of AFMC, of which AFRL is a part. AFMC headquarters is also located at Wright-Patterson. The AFMC commander's responsibilities are very broad, including the programs at four product centers, five air logistics centers, three test centers, and two major specialized centers, in addition to AFRL. The AFMC commander has too many other responsibilities to focus on S&T, and without an S&T advocate at the Air Staff level where budget decisions are made, support for S&T has declined substantially.

The Air Force needs to establish a deputy chief of staff, who is also a member of the Air Force Council, with primary responsibility for planning and managing future Air Force scientific and technical resources. Among his duties, which should include all Air Force technical activities from research through initial production and maintenance, this officer should be the advocate for funding science and technology requirements and for modifying and tracking the implementation of S&T requirements to minimize instabilities in S&T and R&D funding (including new production processes), ensure that adequate funding is budgeted annually, and defend against attempts to "raid" S&T or R&D funds to meet short-term budget shortfalls in other areas. Finally, this officer should be responsible for ongoing assessment of the Air Force technical resources both current and future and should present that assessment to the periodic meeting of senior Air Force leadership at Corona. Ongoing assessments should include the following estimates:

- The number of technical personnel necessary now and in the future, including skills, technical specialties, and years of experience;
- The organizational base and operating requirements to

support the Air Force's long-term technology needs, including size, skills, and responsibilities for universities and in-house laboratories performing scientific and technical research, industrial organizations that the Air Force considers essential to maintaining a competitive supplier base, organizations the Air Force intends to sustain as sole suppliers in particular areas, and Federally Funded Research and Development Centers (FFRDCs); and

- The need for R&D facilities, such as wind tunnel facilities and test ranges, necessary for transitioning technology capabilities and meeting research requirements. These needs must be closely coordinated with the other services and with the National Aeronautics and Space Administration.

The committee realizes that making this assessment will be difficult because of uncertainties about future operational needs, technical advances, and funding. Nevertheless, if the assessment emphasizes flexibility and planning for change, it could help the Air Force make long-term decisions and decrease the likelihood of future crises in technical support. Stability and a sense of direction in the S&T sector of the Air Force would go a long way toward helping the Air Force use its limited resources effectively to support the defense aerospace industry.

The committee is aware that Goldwater-Nichols defines the relationship between the Secretary's Office and the Chief's Office on technical matters. The Secretary's responsibilities are clear, but this should not mean the military does not have very strong concerns about and influence on the technical resources of the Air Force and does not have the need for oversight. The committee believes that the Air Force, both civilian and military, must pay more focused attention to its technical resources if it is to continue to get the best weaponry.

If Goldwater-Nichols or other constraints make a DCS position unworkable, the role and responsibility recommended could be assigned in other ways. The committee understands, for example, that the Navy has established a position for a two-star Director of Test & Evaluation and Technology Requirements (N091) who reports directly to the Chief of Naval Operations (CNO) and has somewhat similar functions. He serves as principal interface between the CNO and the Assistant Secretary of the Navy for Research, Development and Acquisition on RDT&E; Resource Sponsor for Navy S&T (6.1, 6.2, 6.3A) investments; Resource Sponsor for RDT&E Field activities; and Appropriations Sponsor for CNO RDT&E, Navy (RDT&E,N) funding. In any event, the committee believes that high-level attention is needed to get the best result.

3

Work Force Issues

The committee's meetings with senior industry executives indicate that the defense industry is able to hire adequate numbers of engineers at the current work level and in all likelihood will continue to be able to do so in the near future. The real concern they expressed is whether the quality of these engineers will deteriorate (LMAC, 2000; LMSS, 2000; Northrop Grumman, 2000). The defense industry was attractive to engineers during the Cold War. Since then, large cutbacks in Air Force programs and uncertainties about the future have made it less attractive to the engineering profession. At the same time, the need for engineers in other economic sectors has increased, particularly in the areas of information systems and biotechnology, providing engineers with many new opportunities. Nevertheless, the defense industry can still compete successfully for high-quality workers if the Air Force makes the efforts needed to encourage and support them. Fortunately, the Air Force programs involve some of the most sophisticated, cutting-edge technologies and will continue to do so. Therefore, if working conditions can be improved, the Air Force should continue to attract high-quality people.

INDUSTRIAL TALENT BASE

The committee's concerns about prospects for a talented, well-educated, highly motivated, and appropriately experienced defense aerospace science and engineering work force were based on four factors (LMAC, 2000; LMSS, 2000; Kennedy and Lorell, 2000; Shelton, 2000).

First, recent reductions in defense projects combined with reductions in hiring have changed the age-experience composition of the defense aerospace work force.

Second, the reduction in projects means a decline in new starts, hence a decline in opportunities for design experience. This leads highly qualified technical workers to consider the defense aerospace field less desirable. The lack of opportunities makes it much more difficult to attract and retain top talent and to build and maintain the necessary experience base.

Third, the overall decline in Department of Defense (DoD) funding has increased unit weapon system cost; therefore, scientists and engineers will work on even fewer projects in their lifetimes and thus will have less experience across a broad spectrum of technologies.

Fourth, the increase in unit cost could lead to gaps of years, perhaps, between the development of new systems; in the interim, specifically trained and experienced workers may be lost.

Depth of Experience

The change in the age-experience composition of the work force occasioned by the decrease in defense spending raises serious questions about mentoring and the generational passing on of knowledge in the industry. One immediate effect is that older employees who qualify for early retirement may elect to retire because they see fewer opportunities ahead for interesting work (see Figures 3-1 and 3-2). Meanwhile, younger engineers may leave for what they perceive to be better opportunities to learn and gain experience elsewhere. In addition, the short-term result would be that the work force is predominantly middle-aged. As time passes with no new significant hiring, the work force will become disproportionately composed of older, more experienced employees. This has occurred in the aerospace industry and in government over the past 15 years, during which time the number of engineers aged 25 to 34 has fallen from 27 to 17 percent of the work force. In the space sector, only 7 percent of the engineers are under age 30 (CNN, 2000). At Lockheed Martin Aeronautical Company, for example, new hires, which started to decline in the early 1980s, dropped to almost zero in the 1990s (LMAC, 2000). If this trend continues, as experienced workers age and retire, their knowledge and expertise will be lost. The number of available experi-

FIGURE 3-1 Number of engineers who left Boeing in 1999. SOURCE: Boeing, 2000b.

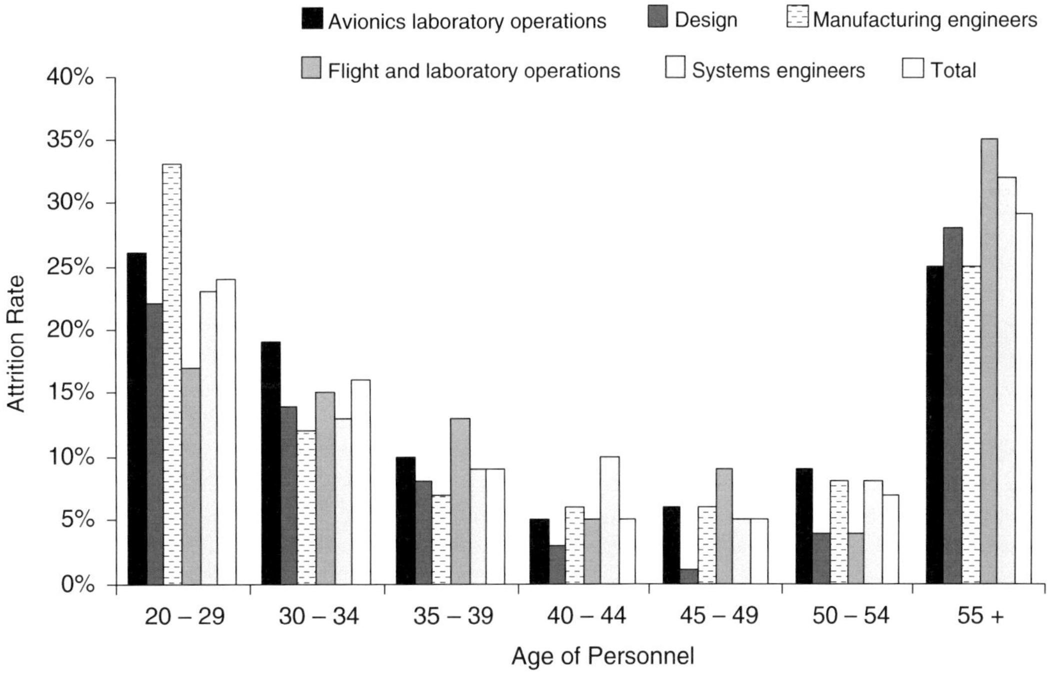

FIGURE 3-2 Rate at which engineers left Boeing in 1999. SOURCE: Boeing, 2000b.

enced leaders will be small, and if the need is there, which seems extremely likely, the gap must be filled by young and inexperienced people.

Many companies are now planning formally arranged mentoring procedures, as well as more training programs for new workers. When new engineers are hired, mentoring and teaming arrangements have to be carefully planned to capture the experience of those about to retire. Nevertheless, experience is lost, as is efficiency, when work tasks involve significant learning curves. For example, since 1982 when the production base started to decline, Raytheon has reduced its work force by 6,000 engineers. The company recently hired 2,200 new engineers, including 500 straight out of colleges and universities. Because Raytheon anticipates that a large percentage of its experienced work force will retire in the next 10 years, it is also bringing retirees back to work on a short-term basis as a stopgap measure. The company has also implemented a mentoring program, popular with both mentors and apprentices, in which experienced people work with new people two days a week. It is also supporting classes during and after working hours (Shelton, 2000).

The Ford Motor Company has also recognized that the company is losing previous practical knowledge when experienced workers leave. To remedy the problem, Ford tried to archive in a database the experience and knowledge for one automobile engine component subassembly. This study showed that at the parts level, about 90 percent of the necessary knowledge and experience could be captured; at the assembly level, this fell to 60 percent; at the system level, it fell to 30 percent. The remainder of the necessary information and experience was potentially lost when employees left (Hastings, 2000).

Historical experiences at the two airplane manufacturing companies of the Lockheed Corporation, the Lockheed California Company and the Lockheed Georgia Company, revealed a need for measuring experience by work association. Because the workloads of the two companies varied, major subassemblies, such as the C-130 wing, were transferred from one company to another. However, a number of cost increases and schedule delays followed. Lockheed found that the most effective solution was to transfer key personnel prior to the move of the hardware. The new team then worked alongside the old one for several weeks or months, developing from observation and conversation the critical information for a smooth-running assembly operation. The significant cost of these temporary reassignments was more than offset by the decrease in start-up costs at the new assembly location.

The experiences at Ford and Lockheed suggest that practical production knowledge cannot be captured in written descriptions suitable for textbooks or manuals. This knowledge includes many judgments about subtle trade-offs learned mostly through experience and best passed on from more experienced employees to their less experienced colleagues. As Ford's experience shows, the more complex the task, the greater is the need for the transfer of knowledge.

Loss of Breadth of Experience

Another issue of concern is the loss of workers with broad experience in working on several different programs. As Table 3-1 shows, the number of fixed-wing, manned, combat aircraft programs has declined steadily since the 1950s. Many aircraft engineers today, even with 10 years of experience, have worked on only one fighter design project, so the breadth of experience is narrowing as older engineers retire.[1] This loss of breadth of experience is at least partially offset by the fact that during the development lifetime of a fighter aircraft it will go through several complete avionics system designs due to the rapid change of avionics weapons technology. The F-16 and F/A-18 are examples; the F-22 will be another.

TABLE 3-1 Decline in Fixed-Wing, Manned, Combat Aircraft Programs

Decade	Number of New Aircraft "Starts"
1950s	46
1960s	15
1970s	12
1980s	7
1990s	4
2000s	1 (to date)

SOURCE: Kennedy and Lorell, 2000.

Compounding the problem of fewer procurement dollars has been the increased cost per unit (in constant year dollars) due to inefficient procurement rates. Procurement dollars can no longer buy as many weapons as they did in the past. Therefore, the number of new systems that can be developed and procured even for the same budget in constant year dollars has decreased. Data for military aircraft show that after adjusting for inflation, the average procurement cost of an aircraft in 1993 was between five and six times as much as it was in 1973. Data show similar trends for other DoD-procured advanced technology systems, such as helicopters, ships, and tanks (GAO, 1997). The increase in unit cost is partly a function of a high-technology approach to the incorporation of very sophisticated, high-cost technologies, such as advanced avionics, night and all-weather capability, precision munitions, and stealth characteristics, into new weapon systems. DoD adopted this approach to take advantage of U.S. world leadership in technical areas and to minimize potential U.S. casualties and collateral civilian damage in time of war.

In extreme cases, the increase in unit cost could lead to gaps of years before the development of a new system. In these cases, development teams would have to be maintained during the hiatus or teams would be disbanded and formed again. If teams are maintained, they must have something valuable to do or they will lose people and skills. In either scenario, costs would be increased and group experience would be lost.

Reconstituting an experienced design organization may take as long as 10 to 12 years. If a competitor is already up to

[1]This problem has also arisen in the nondefense aerospace industry. Many more U.S. commercial designs were developed in the 1940s, 1950s, and 1960s.

speed, playing catch-up may be difficult. On the other hand, a new organization using new technology may be able to leapfrog an old organization. A new organization of capable people can sometimes be better than an old organization set in its ways, but this cannot be counted on. If this time delay is unacceptable, then technical experience must be preserved. According to Etter (2000), "By the time you're worried, it's too late."

RAND looked into the question of the cost of a work lull between major programs, specifically in the development of the Joint Strike Fighter (JSF). In its study of this problem, RAND's Project AIR FORCE defined a minimum core work force to sustain a capability for producing fixed-wing air vehicles as 2,000 people (1,000 specialists in airframe and air-vehicle integration, 500 specialists in avionics, and 500 specialists in propulsion and other areas). The cost of sustaining this core team would be about $500 million per year ($250,000 per year per engineer, plus the cost of facilities, prototyping, materials, support, etc.). Full-scale engineering and manufacturing development (EMD) would cost about four times as much, or $2 billion per year. If a core team were lost during a long hiatus, the report estimated that it would take seven years and cost $3.5 billion to reconstitute the team. The costs of a hiatus, or layoffs and recalls for that matter, include one to two years for hiring and training new workers (even if they are brought in from other firms) and inferior design work and judgment in risk and performance trade-offs (Kennedy and Lorell, 2000).

It is difficult to weigh the costs and benefits of reconstituting a work team as opposed to sustaining an experienced technical team. One consideration is that the military services, and the Air Force in particular, will have to be more flexible than in the past as warfare shifts from traditional platforms to a combination of high-technology platforms and smaller, innovative, new technologies.

> We must change our approach as threats change. We must deal with the new asymmetric threats such as biological warfare and rogue national ballistic missiles. Buying more traditional weapon platforms (like ships, planes, and tanks) will not be sufficient and may be the wrong response. We must be able to take advantage of the rapid improvements in technology, the majority of which are coming from non-traditional defense suppliers who are developing technology for strictly commercial markets. (DSB, 1999)

The mission for the Air Force is changing in ways that are not yet completely understood. The shift from an air force to an aerospace force and an expeditionary force will require technology systems that may not yet be envisioned.

Despite the difficulties of attracting qualified people to work in defense, the committee believes the industry has enough capable people to carry out current Air Force programs and should be able to attract qualified people in the future to develop and build the systems that are now planned. The Air Force is still a major source and sponsor of challenging aerospace research. In addition, the very large reserve of technical talent elsewhere in the national economy could be brought to bear if an emergency arises, perhaps not in a period of months, but certainly in a relatively few years.

The Air Force and industry can, however, do more to ensure that the highest-quality technical people will be available by establishing conditions that will attract and retain them in Air Force research, development, test, and evaluation (RDT&E) and procurement programs.

Level of Program Opportunity

Opportunities for innovative design work that could attract new engineers and ensure broad program experience may be provided by technologies outside of the traditional defense platform programs.

Based on the committee's cursory survey of acquisition programs, between 2000 and 2017, 10 of the top 20 DoD acquisition programs will be aerospace programs, including JSF, the F-22, the F/A-18E/F, a new bomber (the Future Strike Aircraft, or FSA), the Comanche attack helicopter, the V-22, the C-17, the National Missile Defense (NMD) Program, the C-130J, and the KC-XX tanker replacement for the Air Force (Thompson, 2000). These programs are of different scales and in different stages of the development cycle.

Programs in Production

Boeing is producing the F-15E, F/A-18E/F, C-17, and the V-22. Boeing is continuing to build F-15Es since the F-22 is replacing only the F-15C (Boeing, 2001a). The Navy plans to buy at least 548 F/A-18E/F aircraft through FY10, and Congress has approved the construction of 222 aircraft through FY04 (Boeing, 2000a). Plans call for the procurement of 120 C-17s through FY04 (Boeing, 2001b). If technical difficulties with the V-22 program can be resolved, the V-22 will represent a sizable amount of business for Boeing and its partner, Bell Helicopter Textron, Inc. The Air Force has committed to buying 50 V-22s, the Navy 48, and the Marine Corps 360 (USAF, 2000a; USN, 1999).

Lockheed Martin currently has two major aircraft production programs, the F-16 and the C-130J. The Air Force has expressed a need for 70 new F-16s, and 30 have been budgeted through FY04. Lockheed also has contracts to sell 80 Block 60 aircraft to the United Arab Emirates and more than 50 Block 50+ aircraft to Greece and Israel (LM, 2000). On December 22, 2000, Lockheed Martin received a contract for $734.5 million to build 12 C-130Js through FY06 (OASD (PA), 2000).

Programs in Preproduction

Work on the F-22, by Lockheed Martin and Boeing is currently in the EMD phase and is scheduled to enter production in FY03. Plans call for 339 aircraft to be built in the entire production run (LM, 2001). To fill out the structure of the Expeditionary Air Force, the Air Force has expressed the desire to increase the number of aircraft to 415 (Wall, 2000).

The JSF is currently in the concept demonstration phase. The cost of the work in this phase, which began in FY96 and will last until sometime in FY01, is $2.2 billion (JSF PO, 2001a). Total production is projected to be 5,000 aircraft, at a total cost of $400 billion; production will continue through FY27 (JSF PO, 2001b; Morrocco, 2001).

Programs in the Conceptual and Development Phase

Work on the unmanned combat air vehicle (UCAV) is well under way. The Defense Advanced Research Projects Agency (DARPA) is running an advanced technology demonstration program valued at $120 million through FY02 (DARPA, 2000). The Navy is also conducting tests to evaluate the suitability of UCAVs for carrier deck operations. Boeing, Lockheed Martin, and Northrop Grumman all have exploratory UCAV programs for the Navy. If the cost of unmanned air vehicles (UAVs) and UCAVs can be kept down, they may offer many development opportunities for high-quality technical personnel.

Boeing and Lockheed Martin are doing conceptual work on the next generation of tanker and airlift aircraft for the Air Force. The replacement of the KC-135 is not expected until the 2013 to 2040 time frame, but it could happen as soon as 2009 according to the Air Force (Boeing, 2001c; Erwin, 2001). Table 3-2 shows Boeing's scope of involvement in the life-cycle support of various programs and systems.

The Air Force X-vehicle technology programs involve some of the most interesting aerospace research in the world. The Air Force is interested in or is pursuing several space X-vehicle technologies. Development of a space maneuver vehicle is the object of the Air Force's X-40A Program, which relies heavily on the results of the National Aeronautics and Space Administration's (NASA's) X-37 reusable launch vehicle (RLV) program. In July, 1999, Boeing received a $173 million, four-year contract for work on the X-37 (David, 2000). The Air Force is also interested in an orbital transfer vehicle and has several experimental satellites in the works; the XSS-10 with a projected launch date of 2001, the XSS-11 with a planned launch date of 2004, and the TechSAT 21 also with a planned launch date of 2004 (Anderson, 2000).

More conceptual systems under consideration by the Air Force include microsatellites, on-orbit maneuver and spacecraft servicing, sensorcraft (a multisensor unmanned air vehicle), space optics and lasers, and miniaturized munitions (Carlson, 2000; Ruck, 2000). If support for these programs and other advanced concepts increases, it may signal an intensification of the shift from airframes to avionics that began after World War II. This would create some concerns about maintaining existing expertise in airframes but would increase future needs for software development, a highly competitive area.

Exciting work is also going on in the civilian sector. RLV technology was being developed as part of NASA's X-37 technology test bed program. The X-37 is currently undergoing drop tests and is scheduled for an orbital autonomous reentry and landing test in 2002. Other NASA programs include the reduced-cost, small-payload technology experiment; the ceramics-for-sharp-leading-edges experiment; and a Hall-effect thruster. The next round of Pathfinder explorations could include a crew escape system demonstrator (Little Joe III); a reusable, first-stage, glide-back demonstrator (Flybac); an International Space Station fast package delivery demonstrator (Fastpac); a space tug-transfer stage to haul cargo between Earth and the Moon; a nuclear precursor vehicle; and an advanced vertical takeoff-vertical landing demonstrator (London, 2000).

TABLE 3-2 Boeing Life-cycle Support

Future Concepts	Emerging	Growth	Mature	Declining
UCAV	JSF	C-17	F-15	B-1
Common support aircraft	Comanche	F/A-18E/F	T-45	B-2
Advanced theater transport	KC-767	F-22	AV-8B	B-52
Blended-wing-body	F/A-18G	V-22	Apache	F/A-18C/D
Canard rotorwing		JDAM	CH-47	Hellfire
Affordable rapid response missile demonstrator		Brimstone	AGM-130	
		C-40	CALCM	
			SLAM-ER	
			C-32	

NOTE: UCAV, unmanned combat air vehicle; JSF, Joint Strike Fighter; JDAM, joint direct attack munition; AGM, air to ground missile; CALCM, conventional air launched cruise missile; and SLAM-ER, standoff land attack missile-expanded response.

SOURCE: Boeing, 2000a.

Barriers to Continued Aerospace Programs

Despite the diversity of potential programs, program uncertainties are a constant worry for defense aerospace prime contractors such as Lockheed Martin, Boeing, Raytheon, and Northrop Grumman. Any slips or the cancellation of either the F-22 or the JSF aircraft programs would have significant near-term consequences for the engineering work force, because of the absence of equivalent programs.[2] Even if the JSF enters the EMD phase, there will be a 16-year gap between FY08 and FY24 before the Future Strike System is planned to enter the EMD phase. Therefore, even if the JSF enters the EMD phase on schedule, the work force faces an uncertain future (Northrop Grumman, 2000).

Attractions of a Career in Defense Aerospace Engineering

Layoffs in the aerospace sector and abundant opportunities for higher salaries or benefits in other sectors of industry may discourage engineers from pursuing careers in aerospace, and defense aerospace in particular. No company told the committee that it was currently unable to carry out its defense aerospace projects because it could not find good people. However, several "early-warning" signs were cited that the defense aerospace sector is becoming less attractive to talented engineers. For example, Raytheon reported that the acceptance rate of offers to its most desirable potential employees, referred to as the "go-getters," has fallen dramatically (Shelton, 2000).

The shortage of software engineers is an acute problem in the commercial and defense aerospace industry. Even if the supply increases with time, the demand will probably still exceed the supply. Advanced aviation systems have a higher software content than traditional systems, and they require the education and training of software-proficient system engineers. High-potential young software engineers often leave for jobs in nondefense industries where pay scales are higher and perceived opportunities are more exciting. On the other hand, this is becoming less of a problem since the economic downturn of the technology sector as a large number of nondefense industries are proving to be unreliable sources of employment and therefore less competition for the defense industry (CNN, 2000).

The growing gap is illustrated by problems with the upgrading and maintenance of avionics software. Recent studies have shown that the technical competence of maintenance personnel is eroding, particularly in the area of software technology, in both the government and the defense industry. Most of the new avionics software (in a modular open systems architecture [MOSA] environment) will be designed by avionics suppliers, but only engineers at governments depots are familiar with legacy equipment. At the same time, government depots are increasingly using industry personnel to compensate for the diminishing in-house capability that is clearly occurring. Ultimately, government and industry will have to work together to solve the software engineer problem.

The increasing complexity of software systems will exacerbate the problem. The Air Force can mitigate the problem somewhat by sharing best practices in software design and maintenance with industry, by encouraging increases in personal productivity to reduce the need for more software engineers, and by exploring ways to consolidate software support. However, to attract new technical personnel, the Air Force may have to offer hiring incentives to narrow the gap between government and industry offerings (NRC, 2001).

The two JSF prototype design teams at Lockheed Martin and Boeing involve people with critical skills that are easily transferable to the commercial sector, such as computational fluid dynamics, thermodynamics, and avionics and systems engineering, who are constantly being recruited. Because of the relatively high attrition rate among younger engineers, Boeing and other companies have initiated programs to improve productivity, thereby lessening demand. The Boeing effort to improve productivity has resulted in 29 percent less effort being put into drawings, 18 percent less effort into manufacturing engineering, 74 percent less effort into tool design, and 21 percent less effort into software work. To retain employees, Boeing has also initiated other changes, such as work at home for software engineers, software training for nonsoftware engineers, and generous tuition reimbursement programs (Boeing, 2000b).

If the Air Force wants good people to work on its projects, it must provide attractive working conditions, job stability, a competitive salary structure, and efficient program management. A healthy organization, whether in industry or in government, must also provide reasonable physical conditions, benefits, and opportunities for growth and advancement.

ACADEMIC TALENT BASE

The academic sector of the defense aerospace infrastructure is also feeling the budget squeeze. Overall research funding to universities for aerospace research peaked in 1990 at \$106.3 million and averaged only \$78.8 million from 1991 through 1998 (Table 3-3). DoD agency 6.1 and 6.2 funding, NASA funding, and DARPA funding have all deceased.

College and university engineering programs are the major source of technical personnel entering the defense aerospace field. The number of engineering students in aerospace disciplines—aerospace engineering, mechanical engineering, computer science, and electrical engineering—is down. As Table 3-4 shows, substantially fewer B.S. degrees were granted in aerospace engineering in 1998-2000 compared to

[2] Although the recent award of the international UAE F-16 program to Lockheed Martin may mitigate the problem, 40 percent of its design work force is involved in the F-22 and JSF programs (LMAC, 2000).

TABLE 3-3 Funding by Federal Agencies to Universities for Aeronautical and Astronautical Research (in millions of constant FY01 dollars)

	1987	1988	1989	1990	1991	1992	1993	1994	1995	1996	1997	1998	1999
Aeronautical research	68.3	61.2	73.2	85.7	59.5	58.1	53.2	64.5	53.7	48.5	50.3	63.4	72.5
Astronautical research	14.4	18.2	20.2	20.6	22.3	25.6	20.8	21.1	20.1	18.3	18.0	9.0	41.0
Total	82.6	79.4	93.4	106.3	81.8	83.7	74.0	85.6	73.8	66.7	68.3	72.3	113.5

NOTE: Distribution of 6.1 funds: universities, 60-70%; laboratories, 27-37%; federally funded R&D centers, contractors, 3%.
SOURCE: NSF, 2000, 2001.

TABLE 3-4 Aerospace Engineering Degrees Awarded from 1991 to 2000

Type of Degree	1991	1992	1993	1994	1995	1996	1997	1998	1999	2000
Bachelor's degree	2,898	2,915	2,707	2,311	1,789	1,722	1,372	1,291	1,221	1,274
Master's degree	969	1,007	1,080	1,098	875	771	720	638	640	696
Doctorate degree	205	239	207	247	245	262	285	236	195	205

SOURCE: EWC, 2001.

the early 1990s, although the number of doctorates has increased slightly. Most of the work force in applied aerospace have master's degrees; holders of doctorate degrees tend to seek employment in universities and government research laboratories. However, the increase in doctoral degrees may reflect students' decisions to stay in school as long as employment opportunities are low. The number of bachelor's and master's degrees in mechanical engineering and electrical engineering has diminished somewhat from peak levels in the mid-1990s, while the number of doctoral degrees has remained flat or increased slightly. The proportion of bachelor's and master's aerospace engineering degree recipients who are foreign nationals has grown over the past decade by three times and roughly one-and-a-half times, respectively. A slightly decreasing but significant number of doctoral recipients are non-U.S. citizens. Foreign nationals are limited in their opportunities for working in classified defense aerospace, further reducing the number of aerospace engineers available for defense-related aerospace work (EWC, 2001).

The effects of decreasing defense investments are multiplied by the loss of talented students to other fields. Reports from graduates on job opportunities and the quality of work influence faculty perceptions of the attractiveness of aerospace engineering as a career. DoD and the Air Force must maintain their connection to university research to ensure that faculty and students remain engaged in research on technologies important to the military and the Air Force. Students choose to pursue a particular discipline not only because they are interested in the subject, but also because they expect to be employed in the field.

Today, however, the defense aerospace sector is less attractive to engineering graduates for several reasons: financial rewards are lower than those offered by small, high-technology, rapid-growth commercial companies, especially in telecommunications, e-commerce, and biotechnology (although this may be changing); downsizing and continuing mergers and acquisitions have resulted in layoffs, reorganizations, and turmoil in the workplace; and because of the oppressive and intrusive oversight of defense programs, they do not provide as much opportunity for creativity and innovation as private companies.

Although the Air Force can offer unique opportunities in some of the most exciting engineering projects in the world, the message that usually reaches students is about layoffs in the aerospace sector, limited opportunities, and decreasing funds for research. The only program with certain funding, students believe, is for one fighter aircraft, the JSF. The Air Force must overcome the perception that defense aerospace research is no longer important and that opportunities in the field are severely limited (Table 3-5).

Continued investment in university research is crucial to the future defense aerospace infrastructure. The S&T products of university research are often innovative and creative, and investment fuels the pipeline of faculty and students interested in military problems. In fact, the Air Force Office of

TABLE 3-5 Projected Job Growth in Engineering Fields (in thousands)

	Employment		Change in	
Field	1998	2008	Number of Jobs	Percentage
Aerospace engineers	53	58	5	8.8
Chemical engineers	48	53	5	9.5
Civil engineers	195	236	41	20.9
Electrical or electronics engineers	357	450	93	25.9
Industrial engineers	126	142	16	12.8
Materials engineers	20	21	2	9.0
Mechanical engineers	220	256	36	16.4
Nuclear engineers	12	12	1	5.8
Petroleum engineers	12	12	0	(3.6)
All other engineers	415	509	94	22.6
Total	1,462	1,752	290	19.9

NOTE: Numbers may appear not to add or compute correctly. The Bureau of Labor Statistics rounds its numbers to the nearest 1,000. Any apparent errors are attributable to rounding.
SOURCE: BLS, 1999.

Scientific Research has supported 37 people who went on to become Nobel Prize winners.

Academia is also being squeezed by having to share the costs of research, particularly the costs of equipment. Most federally sponsored research is based on cost sharing to ensure that grant recipients are truly dedicated partners in the research enterprise. In the current climate, however, many universities may not be able to provide cost-share funds at the usual level.

Research by universities, faculty, and students supported by U.S. government science and technology (S&T) funds forms an indispensable base for providing young, well-educated technical people for careers in defense technology and Air Force laboratories. The relationship between universities and the Air Force is symbiotic. Universities need S&T funds, and the Air Force needs both the results of S&T and the new engineers. If the Air Force does not shoulder its share of the cost of overhead recovery or facilities, the universities and, ultimately, the Air Force itself will both suffer.

GOVERNMENT TALENT BASE

The other component of the work force is employed by government organizations involved in defense aerospace. These workers have slightly different concerns. Based on site visits to government aerospacc organizations (Appendix A) and other information, the committee identified the following general concerns:

- Substantial downsizing of the technical work force has led to a loss of technical expertise and experience.
- Government rules and regulations on hiring, employment, and retention or firing make it difficult to attract and keep talented individuals.
- Because modernization of facilities and equipment has been deferred, scientists and engineers are no longer working with the best, most modern tools.
- Innovative personnel programs have been initiated to meet these challenges.

Hiring Constraints

The Air Force Materiel Command (AFMC), which employs about 90,000 people at 13 sites, has lost 35 percent of its civilian technical work force since 1989 (personal communication, K. Compton, Public Affairs, Air Force Materiel Command, February 22, 2001) and 45 percent of its military technical work force since 1994 (personal communication, K. Compton, Public Affairs, Air Force Materiel Command, March 21, 2001). Although a slight increase is projected for 2001 and 2002, AFMC has had difficulty attracting new employees, partly because of government hiring regulations. As a result, the work force is skewed toward older workers (Stewart, 2000). Similar reductions have been made at NASA laboratories. At Langley Research Center, for example, the full-time employment ceiling was lowered from 4,000 in the 1970s to 3,000 and is about to be lowered again (Creedon, 2000).

The Air Force Research Laboratory (AFRL) expects to lose 25 to 30 percent of its people in the next five to seven years, an estimated 25 percent loss in the knowledge base, some of it in obsolete and older technologies (Hastings, 2000). It, too, cannot replace its losses by direct hiring because of low salaries and hiring delays related to government constraints. The mean salary difference between scientists and engineers at AFRL and comparable workers at five different government-owned, contractor-operated (GOCO) facilities was $20,000, with the greatest shortfalls at higher-level positions. Examples of hiring delays at AFRL include a number of cases at one facility where delays ranged from six months to more than a year. The average time to hire at another AFRL facility was five months after the candidate was identified. In one instance, five top candidates were lost because of the hiring delay (CSAF, 1999). Government organizations simply cannot maintain a world-class in-house science and engineering capability as long as they are required to operate entirely within U.S. civil service policies and procedures. It should be noted that the Defense Science Board (DSB) Task Force on Human Resources and Strategy found similar situations in the course of its in-depth review of trends and opportunities to improve the ability of DoD to attract and retain critical personnel. The DSB Task Force recommended that necessary legislation be enacted to transfer authority for the DoD civilian work force from the Office of Personnel Management to the Secretary of Defense (DSB, 2000a). It is the opinion of the committee that this recommendation has considerable merit and should receive serious consideration.

To fill out its work force, AFRL plans to augment its cadre of career civil service and military personnel with an equal number of university, nonprofit, and industry personnel. The industry-academia personnel would have limited-term appointments of four to six years. The objective of this arrangement is to make the work force more "agile" in responding to technical and research needs; to improve government personnel management processes; and to mix government staff, which provides continuity and corporate memory, with top industry-academic talent, which provides a different perspective and broader exposure. The initiative takes advantage of newly streamlined hiring authority and other special arrangements (Paul, 2000).

Since World War II, the government has established special long-term contracts with a number of Federally Funded Research and Development Centers (FFRDCs), which have close working relationships with the government but have the flexibility to operate under private-sector rules. The Jet Propulsion Laboratory (JPL), an FFRDC managed by the California Institute of Technology for NASA, is a good example. Its employment structure is not constrained by civil service rules, and it has been relatively successful in recruiting and retaining technical people. According to a spokesperson for JPL, work on space technologies continues to enjoy a mystique and popularity among engineers, and the acceptance rate of employment offers at JPL is between 85 and 90 percent. Despite this success, JPL has experienced a decline in systems engineering experience and a somewhat higher turnover rate among communications and computer specialists who have been lured away to commercial companies, including entertainment firms and software start-up companies (Stone, 2000).

DARPA has initiated a five-year program, Experimental Personnel Management for Technology Workers, to increase its flexibility in hiring 20 "eminent experts in science and engineering" for R&D projects (Seffers, 2000). The program allows DARPA to cut the time of the hiring process from several months to about three weeks. Employees hired under the program are limited to a maximum rate of pay, just as other federal workers are, but they are not assigned pay grades, pay bands, or steps, and initial salaries are negotiable up to the maximum level.

Although organizations throughout DoD are seeking ways to overcome the personnel constraints imposed by government policies and procedures, these initiatives are piecemeal attempts and do not represent an across-the-board effort to improve the government's hiring of technical manpower.

Military Technical Personnel

The committee recognizes the value of having military personnel in the S&T and acquisition communities. The committee did not devote significant time to addressing the status of military technical personnel; however, DoD's policy of assigning uniformed personnel to S&T activities, both in laboratories and in technical oversight positions, is considered valuable to the overall attainment of service goals.

A significant source of Air Force officers with advanced technical degrees is the Air Force Institute of Technology (AFIT) located at Wright-Patterson Air Force Base (AFB). During the1990s, the Air Force contemplated closing the inhouse school at Wright-Patterson. A key factor during this consideration was cost. Congressional concern arose that was reinforced by the trend toward declining Air Force S&T work force size. The Air Force decided not to close AFIT; however, AFIT remains a congressional interest item. The committee supports the decision to maintain AFIT as a major source of technical competence in the uniformed ranks. The committee strongly supports technical education for Air Force personnel at both AFIT and civilian universities. The appropriate balance between them is beyond the scope of this study.

This military S&T work force has been experiencing problems similar to those experienced by the civilian work force. Young, highly motivated officers with advanced scientific and engineering degrees are affected by the same factors that affect civilian S&T workers, including low morale and plentiful challenges, and opportunities outside the military. In addition, there appears to be a perception among some military officers that S&T assignments provide limited career opportunities or are even detrimental to their careers. For example, only nine current Air Force general officers have ever served a tour in an Air Force laboratory (CSAF, 1999). As a result, the number of officers seeking such assignments has dwindled. In 1999, only half of the allocated positions for uniformed personnel at AFRL were filled (CSAF, 1999). This is a telling statistic about the perceived importance of S&T by Air Force officers regarding their career development.

The Air Force is a highly technical organization. It matters that Air Force leadership has the technical training and experience to understand and guide their core technologies.

Outsourcing

Another result of the constrained work force at government facilities is a reduction in in-house S&T and an increase in contracting out the work while maintaining a management function. In the past, employees in government laboratories were personally involved in R&D, as well as in managing contracted-out work. Their own research provided a valuable background for their informed supervision of contract research. As the level of in-house research falls (e.g., the airborne laser [ABL]), government researchers are losing this valuable experience. If this trend continues, government contract monitors will have no R&D experience, which could undermine the effectiveness of contract management. In addition, top-quality people are not likely accept a job that only promises management of others' R&D.

Inflexible civil service regulations, hiring practices, and employment conditions, as well as salary realization, have seriously impeded efforts to attract and retain high-quality technical civilian personnel within the Air Force, particularly in a laboratory environment. The resulting degradation of government research talent has caused more government research to be contracted out to industry and university laboratories. The people who remain in government laboratories are spending more of their time as contract monitors than as researchers. Under existing rules and in the present business climate, the government has difficulty maintaining a highly qualified technical work force, except in a few instances where progressive personnel programs have been allowed on a pilot basis.

Allocation of Funds

DoD personnel responsible for funding and overseeing programs face serious and increasingly difficult challenges. With the decline in defense investment in S&T and R&D, policies and programs must be organized and executed effectively. Efficient management and allocation of funding are critical factors in technology advancement, especially when resources are diminishing. The lack of industry experience in the Air Force senior leadership is a significant problem. These leaders must create policy; must manage, craft, and execute programs; and must be "smart buyers" for the Air Force to continue to generate advanced technologies. The committee examined, for example, 70 biographies of senior Air Force civilians involved in funding and overseeing programs. Of the 70, only 10 percent had at least one science or engineering degree and had worked for an aerospace manufacturing firm; 43 percent had a technical education only; 3 percent had industry experience only; and 44 percent had none of these (USAF, 2000b).

4

Financial Health of the Aerospace Industry

The aerospace industry has undergone significant restructuring in the last 20 years. A number of recent defense studies examining the implications of these changes reflect a broad consensus in the aerospace community on the characterization of the defense aerospace industrial base. The agreement on the general characterization of the aerospace environment is also reflected in highly congruent recommendations to the Department of Defense (DoD) for responding to the changes in the aerospace environment.

AEROSPACE INDUSTRY ENVIRONMENT

The aerospace industry restructuring has changed the financial metrics of the industry. In the course of this study, several studies of the defense industrial environment were conducted by other organizations (e.g., DSB, 2000b and Harbison et al., 2000). Although these studies covered the entire U.S. defense industrial base, their findings and recommendations are directly applicable to the aerospace sector.

One result of the consolidation in the defense industry has been that most defense companies now have a customer base that not only spans the entire spectrum of defense products, but also includes commercial products, which yield higher margin for less trouble. Large companies, in particular, continue to diversify to compensate for low return on investment (ROI) from government contracts, and only part of the aerospace sector produces defense products. The defense aerospace industry can be characterized as an industry in transition with many companies facing challenging problems (DSB, 2000b):

- Because of strong competition and stringent export controls, opportunities for growth are limited.
- Profitability, already low compared to other industries, has declined. In addition, companies that encounter problems on major programs face potentially further reduced profits.
- Cash flow, which has traditionally been a strength of the defense industry, has declined for most companies.
- As a result of consolidations, some companies have added to their debt, creating higher debt-equity ratios, which have resulted in lower credit ratings.
- Market capitalization by defense companies has declined more than those for most "old-economy" companies.
- Innovative research and development (R&D) has been reduced; funding by DoD for R&D has been flat. Funding for innovative R&D is down 50 percent from the mid-1980s and is increasingly focused on supporting ongoing programs rather than on breakthrough technologies.
- In an era of few large production programs, the Cold War approach of "getting well on production," that is, making up for research expenses in the production phase of a program, is no longer viable.
- Key personnel are leaving or retiring, and retaining and recruiting new high-quality technical and management people are difficult.

In *U.S. Defense Industry Under Siege: An Agenda for Change,* the industry was characterized as being at a crossroads. Although industry financial metrics have improved since the publication of that report, the underlying conditions have not changed (Harbison et al., 2000).

> The underlying health of the industry is seriously suspect. As we entered the new millennium, the industry's combined operating profitability has declined from 9.2 percent in 1996 to 7.7 percent in 1999, and the industry's collective interest coverage ratio has fallen to 2.7 times in 1999 from 7.1 times in 1995; debt ratings have fallen to almost junk bond levels, and the industry's market capitalization is down 33 percent from \$100.1 billion in January 1997 to \$66.7 billion today.

At the Defense Reform 2001 Conference organized by the American Institute of Aeronautics and Astronautics, the industry environment was discussed by top industry and government officials, who called for the following changes (Velocci, 2001):

- An immediate increase in the progress payment system from the current 75 percent, which constrains cash flow, to 85 to 90 percent;
- Changing the export control process, which inadvertently penalizes U.S. companies and enables potential adversaries to acquire restricted military technologies from other sources;
- Making it easier to use commercial technologies; and
- Making it easier to retain design teams.

The studies discussed so far reflect the broad consensus of the defense industrial community. The results of the committee's own investigations substantiated their findings and recommendations. The recommendations in these studies are summarized below:

- The partnership between DoD and industry must be strengthened.
- Programs and funding must be stabilized.
- Creative incentives must be provided for the industrial base to rationalize capacity.
- Single providers must be carefully selected and managed.
- The spirit of innovation must be encouraged.
- Industry concerns must be considered in the DoD acquisition process.
- Industry metrics must be better understood.
- Export control processes must be streamlined.
- Human resources issues must be addressed.

INFLUENCE OF THE U.S. DEPARTMENT OF DEFENSE ON THE AEROSPACE INDUSTRY INFRASTRUCTURE

Even though the defense industry has been dramatically consolidated since the end of the Cold War and the relationship between the industry and DoD has changed dramatically, the fundamental policies of DoD have not changed. DoD's share in the aerospace market is shrinking as a result of an increase in nondefense sales and a decrease in DoD procurements. In 1989, DoD accounted for 51 percent of aerospace sales in the United States (see Table 4-1). Since then, DoD's spending on aerospace items has returned to pre-Reagan levels. In 1999, DoD accounted for only 30 percent of aerospace sales (AIA, 2000, 2001a).

In 1977, 15 percent of the national investment in R&D was spent on aerospace. Today, as more and more R&D dollars are spent in other fields (e.g., pharmaceuticals, information systems, biotechnology), the proportion of investment in aerospace has dropped to less than 7 percent (NSF, 2001). The full extent of these influences is shown in Figure 4-1.

TABLE 4-1 U.S. Aerospace Industry Sales in the United States (in millions of constant FY01 dollars)

Year	Total Sales	Sales to DoD	DoD's Percentage of the Total
1984	141,175	72,661	51
1985	159,825	88,010	55
1986	170,211	94,835	56
1987	170,182	95,631	56
1988	170,125	91,071	54
1989	170,797	86,719	51
1990	180,600	81,315	45
1991	178,340	72,139	40
1992	173,516	65,357	38
1993	149,791	57,173	38
1994	131,122	51,941	40
1995	124,273	48,888	39
1996	130,829	47,489	36
1997	144,082	47,854	33
1998	159,534	46,286	29
1999	159,405	47,559	30

SOURCE: AIA, 2000, 2001a,b.

In addition, the U.S. share in the world aerospace market declined from 70 percent in the mid-1980s to 55 percent in 1997 (NRC, 1999). In constant FY01 dollars, it went from $160 billion in 1985 to $146 billion in 1997, a 9 percent decrease (AIA, 2001b).

The environment in the commercial aerospace sector is being shaped by a rapidly expanding economy and by strong free-market forces. Growth in revenue and earnings is strong, the financial markets are supportive, and market capitalization for many industries has never been higher. The aerospace industry is now competing in a market with many technological opportunities and growing financial returns.

DoD is a monopsony (i.e., the only buyer) in the defense aerospace sector. A monopsonistic industry operates much differently than a competitive industry because the single customer ultimately provides the resources that attract workers and capital. There are few, if any, perfectly free markets anywhere with many suppliers and many buyers, perfect information, and no applied restraints. The DoD as a monopsony, or single buyer, for the defense industry cannot be said to operate in anything like a free market. This, however, does not mean that there is no competition, just that the competitions are established and controlled by DoD. The DoD has widely varying relationships with its suppliers, ranging from open competitions to what are essentially permanent single sources and everything in between. Since DoD sets the rules, it is responsible for the effects of these rules on its supplier base whether it recognizes this explicitly or not. Therefore, DoD is ultimately

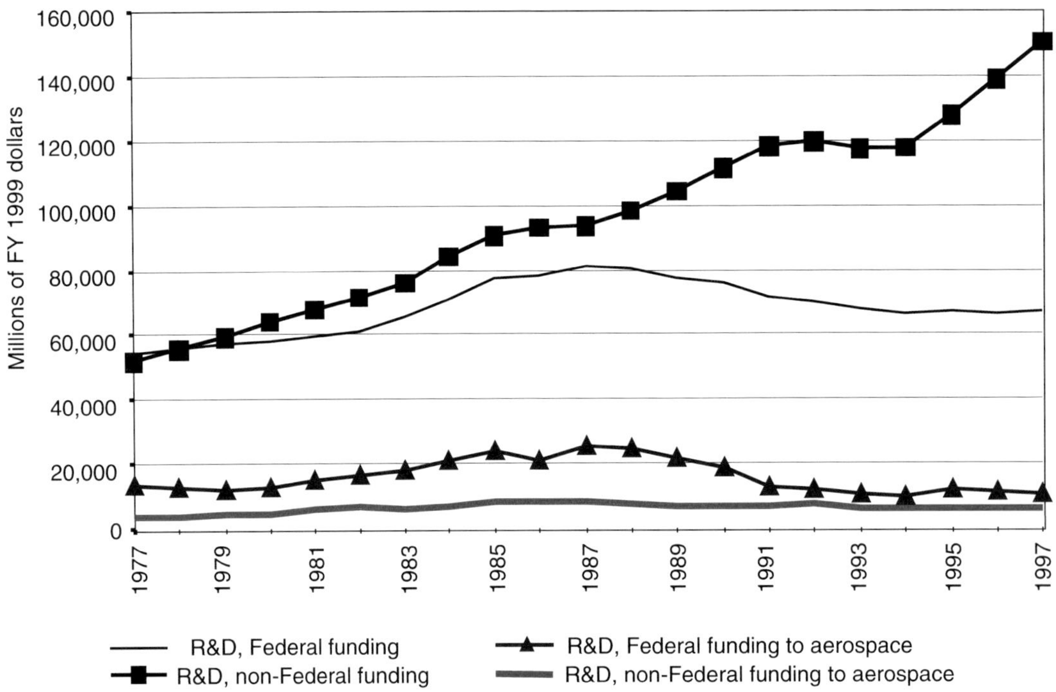

FIGURE 4-1 Funding for R&D by source. SOURCE: Douglass, 2000a.

responsible for the health of the defense aerospace industry, although not for the health of any particular company.

Each of the services is the primary customer for a segment of the defense industry. As the primary customer for the defense sector of the aerospace industry, the Air Force has a significant share of responsibility for its viability. Therefore, the way in which the Air Force uses its resources will have a major impact on that sector of the defense aerospace infrastructure. The committee understands that this is well recognized within DoD but that there seems to be no established mechanism for determining and including the effects on the Air Force's technical resources when decisions are being made about the competitive conditions for individual programs or for the Air Force program as a whole. The committee therefore recommends that this task be a responsibility of the recommended deputy chief of staff (DCS). Air Force management and budget decisions affect fundamental elements of the defense aerospace sector (e.g., program risk and stability, profit margins, proprietary ownership, investment in technology development, authority to transfer technology into commercial applications). Given the vulnerability of the defense aerospace sector, Air Force management would be well advised to take industry concerns into consideration in the development of its budget and management policies. In short, the Air Force has a responsibility and a crucial interest in making decisions that will ensure the health of this industry and that Air Force needs are met.

5

Policy, Administration, and Regulation

Government policies can mitigate or exacerbate the negative effects of budget cuts and program instabilities on the aerospace infrastructure. U.S. government and Department of Defense (DoD) policies have a double impact on the allocation and administration of program funds. First, they directly impact how much technical progress will be made for the taxpayer dollars spent. Second, they have an indirect impact on the quality of the defense aerospace technical work force—on both experience levels and the attractiveness of working in defense aerospace. Thus, policy makers must take into consideration the direct and indirect effects of policy and budget decisions.

ACQUISITION CYCLES

Reducing Cycle Times

The committee identified the effects of long acquisition times (i.e., the time between program initiation and initial fielding) on programs and evaluated the implications for the defense aerospace infrastructure. A 1986 Packard Commission report stated, "An unreasonably long acquisition cycle—10 to 15 years for our major weapon systems is a central problem from which most other acquisition problems stem" (CDM, 1986). The report noted that the long acquisition cycle also leads to unnecessarily high development costs (CDM, 1986). Today, more than 15 years later, the problem continues. In contrast, development cycle times in the commercial sector driven by competitive pressures have fallen dramatically. As of year 2000, the commercial aircraft industry had cut cycle time from 8 to 10 years to 5 years; the spacecraft industry had cut its development cycle time from 8 years to as short as 18 months (this reduction reflects, in many cases, modifications of existing platform designs as opposed to the development of new platforms); and the consumer electronics industry, from 2 years to 6 months. These reductions demonstrate that with a concentrated effort, cycle times can be cut in half or better (Hastings, 2000).

Long cycle times not only impact the quality of the products delivered to our warfighters (e.g., products are based on requirements more than 10 years old, and technologies are frequently obsolete and sometimes out of production), but also have a significant impact on the defense infrastructure that designs, develops, and produces these products.

First, the relationship between the development cost for a defense system and its development schedule is not linear (McNutt, 1998). Second, long acquisition cycles are major contributors to program instability, the most disruptive of which is the higher rate of cost increases. In 1996, the Air Force established the Lean Aerospace Initiative (LAI) to address this and related problems. Surveys of more than 100 government program managers and 80 contractor managers revealed that program cost growth attributable to budget changes alone averages just over 2 percent, *year after year* (LAI, 1996a, 1996b). The cumulative effect for an 11-year acquisition cycle is significant. In addition, cost increases in one program generally mean that budgets elsewhere must be reduced to make up the difference.

Cancellation is the ultimate form of instability, and the probability of cancellation for a defense program is just over 4 percent per year for each year the program is in development (Augustine, 1996). In a rapidly changing environment, programs based on 10- or 12-year-old requirements become increasingly vulnerable to cancellation. Unfortunately, programs canceled late in the development process result in zero return on very large investments. For example, four generations of Army air defense systems were terminated at an investment of $6.7 billion (Johnson, 1995). In addition to these huge losses, program cancellations are a major contributor to the difficulty of attracting qualified personnel.

Changes in program direction also occur frequently as the result of changes in leadership and changes in defense program annual guidance issued during the annual budget cycle. During an 11-year acquisition cycle, the leadership at each level, from the program director to the secretary of defense, typically turns over four to eight times; the most frequent turnovers occur at the service and defense acquisition executive levels (Eash, 1998). These changes are reflected in shifting priorities and associated budget cuts and redirection (LAI, 1996a, 1996b). Budget cuts and program redirection often result in programs being stretched out as well. Of the top 20 acquisition programs now in place or planned, 6 had delays due to funding shortfalls, 5 had production cuts or stretches, and 3 had uncertain funding or were unfunded. The cost and commonality goals of the Joint Strike Fighter (JSF) have been characterized as unattainable. The F-22 production goal has been cut four times in eight years, unit cost is rising (partly as a result of production cuts), and congressional resistance to continuing the program is increasing. The production goal for the F/A-18E/F was cut by 45 percent in the Quadrennial Defense Review (QDR), and some Marine Corps resistance to continuing the program has been encountered. The Future Strike Aircraft (FSA) is currently unfunded and will not be produced prior to 2030. Numerous restructurings of the Comanche program have been forced by funding shortfalls. Production of the C-130J was delayed due to a funding shortfall. The KC-XX is as yet unfunded (Thompson, 2000).

The benefits of shorter acquisition cycles would be enormous. First, shorter development programs would allow individual workers to participate in more programs and thus increase the breadth and depth of their technical knowledge. Thus, working on Air Force programs would be more attractive. Second, shortening the acquisition cycle would save money that could be used to support more programs. Third, besides enhancing military capability, shorter cycles would reduce gaps between programs and reduce the loss of specifically trained, skilled, and experienced workers. Fourth, shorter cycle times would improve program stability by reducing cost overruns, cancellations, and changes in program direction. Reducing these disruptive changes in programs would lead to sizable cost savings. Improved stability in the workplace would make a career in aerospace more attractive. Finally, shorter cycle times would allow engineers to spend more of their time engineering and less on the busy work related to the acquisition process.

Using Commercial Products and Processes

DoD acquisition policies have made it difficult for defense systems to take advantage of the latest commercial cost benefits and dynamic advances in commercial technologies. This has implications for the defense infrastructure. The size of the defense infrastructure depends on a combination of two factors—weapon subsystems and components that cannot employ commercial solutions and subsystems and components that could use them but do not.

As the use of commercial off-the-shelf (COTS) technologies and products increases, the need for defense-unique development capabilities decreases. A report by the RAND Corporation, *Cheaper, Faster, Better? Commercial Approaches to Weapons Acquisition* (Lorell et al., 2000), includes a case study showing the cost savings of inserting commercial-grade parts in the manufacture of military avionics components on higher volume, automated, dual-use production lines. The report cites the results of a U.S. Air Force program at Wright Laboratory for the development and manufacture of lower-cost modules for fighter and helicopter systems. Taking maximum advantage of commercial parts, the two modules were estimated to cost about 60 percent of the original F-22/RAH-66 baseline cost projection, even though the modules had not been designed for COTS insertion. The program did not permit basic electrical redesign of the modules. Partly because of this restriction, 10 percent of the parts remained military specification (Mil-Spec) and accounted for 50 percent of the module cost. The report concluded that even though commercial parts would have to be screened and possibly made more rugged or repackaged prior to use in military systems, they would often be less expensive than Mil-Spec parts (Lorell et al., 2000). In addition to the technical gains in performance, quality, and supply, the Air Force did not have to fund the development of commercial components.

The uniqueness of the defense infrastructure is related not only to unique technological requirements (e.g., operating in extreme conditions, such as high or low pressure, high or low temperature, the presence of nuclear radiation), but also to different design, development, and production processes and to all of the supporting disciplines. As the defense sector embraces a larger and larger share of commercial processes, the differences between the commercial and defense sectors will also diminish, and therefore the differences in the specialized training, skills, and experience of the work forces will diminish. Thus, the barriers that restrict movement of the work force between these sectors, during both peacetime and periods of crisis, will come down. In areas where the differences are small, commercial industry will provide an abundant source of skilled workers, and the issue of infrastructure maintenance will begin to recede. The challenge, therefore, is first to maximize the use of commercial solutions to satisfy military requirements and second, if commercial products are not acceptable, to adopt commercial processes and practices wherever practical in the development of defense-unique solutions.

Because many critical technologies and processes have few if any commercial analogues, the use of COTS products and processes can reduce the magnitude of the aerospace infrastructure issue but will not eliminate the need for a dedicated defense aerospace industry.

Reforming DoD's Acquisition Processes

DoD recently commissioned the Defense Science Board (DSB) to review recommendations for financial policy changes in response to the declining financial health of the defense aerospace industry (DSB, 2000b). Based in part on these recommendations, the Pentagon is implementing 27 regulatory and policy initiatives to improve the financial health and reflect the restructuring of the defense industry. The Pentagon will allow defense companies to keep part of the savings from efficiencies and reductions in overhead rather than continuing the current practice of returning the savings to the Pentagon. This will provide defense companies a greater opportunity to profit from R&D programs. Defense companies have traditionally taken greater financial risks in the R&D phase on the chance of larger profits in the production phase. The Pentagon also plans to increase multiyear contracts to stabilize production and lower prices. In addition, the Pentagon will increase its R&D spending by 10 percent annually over planned levels of $38 billion in FY02 and $37 billion in FY03 and FY04 (Capaccio, 2000).

In the last two years, DoD has begun to address the problem of acquisition cycle time in earnest. It has established a goal of reducing the average acquisition cycle time by 50 percent for all major defense acquisition programs (relative to a baseline of 11 years) started in FY99 or later (OUSD (AT&L), 2000). Reductions will be based not on reductions in the schedules for existing processes and activities, but on the development of a new way of doing business, including new processes, new activities, and new organizational structures and relationships. The objective is to develop a more effective, more efficient, and more time-sensitive approach. The focus on cycle time provides the vision and objective for implementation of a "lean" philosophy and serves as a metric of success in achieving that objective. Industry's focus must be on improving design, development, and production processes and organizations. DoD's focus must be on the establishment of policies and procedures that will enable this evolution.

The committee strongly supports this initiative. However, the committee cautions that extensive changes to procedures, concepts, and practices require careful deliberation. Changes should be made only after scrutiny by those who have had extensive experience in the processes and operations being revised. A headlong rush to embrace "faster, better, cheaper" can lead to some very undesirable and unexpected consequences.

The report on the National Aeronautics and Space Administration's (NASA's) failed Mars Climate Orbiter identified a "lack of identification of acceptable risk by the operations team in the context of the 'Faster, Better, Cheaper' philosophy" as a causative factor in problems with the program management (NASA, 2000). In short, changes made to improve cycle acquisition time require the attention and active participation of the highest and most experienced levels of senior management in DoD, the Air Force, and industry (Smith and Reinertsen, 1995).

Senior leadership in the Air Force has been visible in support of the cycle time reduction initiative. Both the Vice Chief of Staff and the Assistant Secretary of the Air Force for Acquisition have taken strong advocacy positions (Delaney, 1999; Lyles, 1999).

DoD has already taken some steps to shorten cycle times. For example, Instruction CLCSI 3170.01A of the chairman of the Joint Chiefs of Staff has been modified to suggest that time-phased requirements be used when feasible and to stress evolutionary acquisition and increased technical maturity before acquisition is initiated. The October 23, 2000, revision to DOD 5000-1 significantly emphasizes reducing cycle time in the following ways:

- A rapid, effective transition from science and technology (S&T) to product development;
- Using time-phased operational requirements;
- Requiring demonstration of a technology prior to the start of formal acquisition;
- Placing priority on evolutionary acquisition strategies based on time-phased requirements, proven technologies, and demonstrated manufacturing capabilities; and
- Initiating formal acquisition at or between any of the formal development milestones.

Continued efforts to reduce cycle time will be critical. Even after the current policy changes are fully implemented, much will remain to be done. At Toyota, the leader in implementing the lean philosophy, it took more than three decades for the reforms to permeate internal, supplier, and distributor systems. Toyota continues, even now, to find ways to make significant improvements (Womack and Jones, 1996).

Current DoD reforms are but first steps toward dramatically reducing development time lines, and they will contribute measurably to the efficiency of the defense industry. They will also allow greater flexibility in program design and lead to the development of processes more in tune with industry cycle times. Reform should be considered as a long-term undertaking, however, that will require continued visible support by Air Force leadership and civilian leaders in the Office of the Secretary of Defense.

PRODUCT CYCLE PHASES

Another aspect of the acquisition cycle that affects industry's long-term viability and the health of the infrastructure is the distribution of work across five product (program) phases: (1) concept development, (2) demonstration and validation (essentially prototyping), (3) engineering and manufacturing development (EMD), (4) production, and (5) sustainment. Low industry profits, sometimes even losses, in the first two phases have historically been made up by higher profits in the EMD phase. Today, however, the amount of industry EMD has been reduced, which decreases overall profits.

Each phase of the product cycle requires different engineering skills. Indeed, if one phase were omitted for a period of time, such as the EMD phase on a new aircraft, the engineering teams for that project would have to be disbanded. This would lead to increased costs, either to retain people during the hiatus or to reconstitute a team after the hiatus. Experience has shown that disbanding and reconstituting teams after a substantial period of time (e.g., five years) increases the cost of the subsequent EMD phase by about 35 percent. For a typical military EMD phase of $15 billion dollars, the increase would be $5 billion (M. Kennedy, RAND Corporation, personal communication, February 25, 2001). Industry is concerned that such reconstitution would be very long and expensive. The lesson for the Air Force is clear. The industry cannot survive by working in only one phase of the product cycle.

Engine Component Sector

The aircraft engine component sector of industry has been a notable success. The committee reviewed this case in depth to determine if lessons for other industry sectors could be identified. The review revealed that a key factor in the success is the commonality between critical technologies in military and commercial engines combined with the military's long-standing practice of investing in engine technology programs, such as the Integrated High Performance Turbine Engine Technology (IHPTET) Program. Stable investment has provided the continuity to sustain this sector.

Another aspect of the engine component sector that contributes to its success is the life-limited components and associated repair and replacement business. The lifetimes of engine parts are considerably shorter than the lifetime of the airframe. This creates a continuing revenue and profit stream for engine manufacturers on which investment strategies can be based for technology generation and development for product improvement and replacement. Involvement in the after-market business also enables the engine supplier to become actively involved with products that sustain engineering, thus evening out the workload between product launches.

In addition, the engine industry has benefited from a long-term technology investment program by the military. Investments by industry, the military, and NASA have fueled the technology initiatives in the IHPTET umbrella program on advances in engine technologies and meeting specific military requirements. In appropriate situations, the industry has applied advances in core engine technologies for commercial use. This dual-use arrangement has benefited the military by having the commercial experience feed back into the common military components and has benefited industry by sustaining it between military programs.

In summary, because a unique characteristic of the aircraft engine industry is that defense products can piggyback onto its commercial products, this sector is not a good model for the aerospace industry as a whole. However, this sector does illustrate the importance of sustainment. Long-term investment in technology programs such as IHPTET has enabled this sector to avoid the problems facing other industry sectors such as airframes.

The phases of the acquisition cycle taken together provide a business base that generates the profit and investment that enable firms to grow and prosper. For example, initiating prototype programs for the purpose of maintaining aerospace design teams but without robust EMD and production programs will not sustain the industry, although it might sustain the design teams that may be the "long pole" in reconstructing industry's capability.

RAND has suggested a continuous program of three simultaneous military aircraft prototype programs, each lasting five years, which would require $1 billion per year in current dollars of dedicated funding and could mitigate the problem of loss of experience. Under the prototype plan, workers would gain experience on a new program every five years (Lorell, 1995; Lorell and Levaux, 1998). Similar programs could be used for aircraft engines and avionics. However, engineering talent is a rare commodity in today's highly competitive technology marketplace, and firms are very reluctant to spend their scarce engineering resources for prototype programs without the prospect of profitability during production.

Although the committee recognizes and strongly supports the contributions that prototyping can make to attract, train and retain skilled professional designers, after considering alternatives, the committee concluded that continuing activities from R&D to EMD through production and product support are essential for maintaining all of the skills and team experience necessary for the aerospace industry to produce new aircraft. While prototyping was not addressed in depth by the committee, a separate study of the subject should be considered.

EXPORT LICENSE CONTROLS

Another inhibiting factor in the expansion of aerospace markets is the restrictions imposed by licensing regulations, the International Traffic in Arms Regulations (ITAR), and the Arms Export Control Act. The U.S. government regulates arms exports to ensure that they do not adversely affect national security. The Arms Export Control Act authorizes the president to control the export of military items, which are licensed through the ITAR.

Nevertheless, exports are a major source of production, employment, revenue, and profits for the U.S. aerospace industry. The international component of sales for Lockheed Martin Aeronautics Company is now 70 percent; in the 1980s, international sales accounted for only 30 percent. In fact, international sales sustain the technical base and profits for investments in R&D and modernization (LMAC, 2000).

Foreign aerospace sales and cooperative projects produce a significant revenue stream that can be used to fund R&D.

For example, the Air Force is buying 30 F-16s from Lockheed Martin in the next five years, but Lockheed Martin has also negotiated significant foreign sales, including a large sale to the United Arab Emirates for the Block 60, a more capable F-16 model (LMAC, 2000). Current plans for an Air Force purchase of F-22s from Lockheed Martin call for 339 craft, with potential sales to some allied nations such as Australia (Wall, 2000). If Lockheed Martin wins the JSF contract, it plans to sell 2,800 JSF aircraft domestically, along with significant foreign sales, possibly a Harrier replacement for the British Royal Navy. Lockheed Martin has two significant cooperative programs with the Japanese for the F-2, a version of the F-16, and with the South Korean government for the KTX-2. Korea plans to build as many as 100 aircraft for its own use and to export the aircraft widely.

Export controls penalize American aerospace companies trading in the global market while providing little benefit to national security. In fact, they do the opposite by weakening the U.S. defense industry. Unnecessary regulations control many items that are readily available on the international market, such as advanced computers and encryption codes. In addition, the process by which these regulations are implemented is very complicated, time consuming, and uncertain; each expensive and burdensome paperwork and bureaucratic negotiation requires multiple permissions. The United States has two systems for controlling exports. The first, established under the jurisdiction of the U.S. Department of State, is designed to control the export of military products and technology. This system restricts exports unless there is a national foreign policy or security rationale for exporting the product. The second, established under the jurisdiction of the U.S. Department of Commerce, is designed to control the export of commercial and dual-use products, that is, products that are used in the commercial world but have military applications. Dual-use products are controlled if there is a security reason to do so; commercial products are controlled only where foreign policy sanctions have been invoked.

There is general agreement that this dual system is confusing at best and inefficient and counterproductive at worst. Many items are included in the State Department military product system simply because they have been modified for use on a military product, even if comparable commercial products and technology are widely available in the global marketplace. If such products must be controlled, they should be controlled as dual-use products by the Department of Commerce.

A major problem with the State Department licensing process is that only about 18 staff members in the Office of Defense Trade Controls (ODTC) have been responsible for processing approximately 45,000 licenses in the past several years. If a license is referred to DoD or other agencies for comment, the average licensing time is about 100 days. The problem was exacerbated in 1998 when Congress transferred commercial communications satellites and components from the Department of Commerce system to the State Department because of the alleged leakage of some rocket technology to China. This legislation has imposed additional restrictions, hampering the sale of communications satellite systems and services, sometimes putting U.S. industry, and therefore the Air Force itself, at a disadvantage (Douglass, 2000b).

For the launch of a U.S.-built satellite on a U.S.-built rocket from a U.S. launch facility for a foreign customer, the U.S. company must have an export license, which requires DoD and intelligence reviews and a technical assistance agreement from the State Department. In addition, a technology transfer control plan, an extensive plan that requires Department of State and DoD approval, must often be filed. If the product is valued at more than $50 million, congressional notification and approval are also required. Depending on the type of license, the entire process can take 10 months or longer. According to the Department of Commerce, exports of U.S. communications satellites and components fell by 40 percent as a result of satellites being placed on the State Department's list of controlled munitions in 1999 (Reinsh, 2000).

After meetings between congressional representatives, including the Armed Services committee, the Foreign Relations committee, and the appropriation subcommittees, and industry associations, funding for ODTC has been increased; the review process for satellite exports to the North Atlantic Treaty Organization (NATO) and major non-NATO allies has been expedited; and export process reforms in the Defense Trade and Security Initiative have been ratified. These changes are expected to reduce the backlog, but more will be needed.

Despite export license reform, however, recent legislation, responsive to the alleged unlawful transfer of sensitive missile technology to China, has imposed additional restrictions hampering the sale of communication's satellite systems and services, sometimes putting the U.S. satellite industry at a competitive disadvantage against European systems (Douglass, 2000b). The policy seems to be based on the assumption that the benefits of export sales accrue only to the buying countries and defense industry stockholders. This is simply not the case. Export sales of military items are an important element of the defense of the United States. The perils of exporting the wrong things must be weighed against the real benefit of exporting the right things. The State Department is not necessarily equipped to make this trade-off.

Although it may be difficult to strike a balance between fair trade and national security during periods of military and political change around the world, there are clear indications that the current balance is weighted in favor of restrictions. John W. Douglass, president of the Aerospace Industries Association of America, testified before the Senate Armed Services committee in March 2000, "German irritation with our licensing process is such that managers of DaimlerChrysler Aerospace, or DASA, Germany's largest aerospace company, have been instructed to avoid purchasing American components for defense and space products." Douglass noted that DASA is in the process of merging with French and Spanish companies to form the fourth largest international aerospace and defense company. Restricting U.S. trade leaves room for new European and Asian entrants into component markets currently dominated by the United States (Douglass, 2000b).

Vigorous exports of U.S. aerospace products contributes to the financial health of the aerospace industry, it's breadth and depth of work, the industry's competitiveness in the global economy; and its ability to support national security interests. These national security issues should be given weight along with the security issues that result in extensive trade restrictions.

TEST FACILITIES

The term "test facilities" covers a very broad range of test capabilities used by the aerospace industry and government laboratories. The committee focused specifically on the test facility infrastructure of the "aviation" segment of the aerospace industry.

Full-scale, complete-airframe, static and fatigue test installations and other specialized test facilities can be very large and very complex. Nevertheless, they can be constructed and calibrated prior to the time they are actually needed for testing a particular aircraft design. This cannot be done for wind tunnels used for testing airframes and engines. Although computational fluid dynamics has made impressive strides in the last several decades and can define a best overall configuration, details must still be optimized. Fine-tuning engine inlets, exhaust systems, and wing-fuselage fillet configurations and minimizing interference shock effects at transonic speeds usually require testing in wind tunnels. Free-flight tests are usually more expensive and have a more limited range of test parameters than wind tunnels.

It takes years to build a large, well-equipped wind tunnel test facility and bring it to operational status. The long time interval is not dictated solely by design and construction requirements. Considerable time is also required for running extensive tests to calibrate a new facility. Because wind tunnel testing is required early in the design process of airframes and propulsion systems, facilities must be ready as soon as a decision is made to design and produce flying hardware. Wind tunnels are expensive to build, operate, and maintain.

The use of the nation's wind tunnel facilities has declined significantly. As a result, some facilities have been closed or placed in an inactive state. Boeing is closing one of its tunnels, and several facilities at Arnold Engineering Development Center (AEDC) are becoming inactive because of lack of use (Boeing, 2000b; Heil, 2000). The major national ground test facilities used in DoD programs are at AEDC and NASA centers (Ames, Langley, and Glenn). Flight tests are conducted at a number of Air Force bases (AFBs), principally Edwards AFB and neighboring NASA-Dryden Flight Research Center. The viability and health of these facilities have been of much concern in the last decade or so because many of the major wind tunnel facilities have been in service for more than half a century. The alarm was raised by the failure in 1986 of the pressure shell of the 12-foot tunnel at the NASA-Ames Research Center. The facility has since been repaired, but it took many years before sufficient funds were made available.

A 1988 National Research Council (NRC) study of the NASA facilities, *Review of Aeronautical Wind Tunnel Facilities*, concluded that the NASA test facilities required serious immediate attention in terms of (1) maintenance and upgrading, (2) productivity enhancement, and (3) accommodation of new requirements (NRC, 1988). Although many facilities have been shut down in the intervening years, the more detailed recommendations of the study have generally been followed. The facilities that have been built since 1988 are primarily research facilities rather than testing facilities.

A similar NRC study of AEDC facilities, *Future Aerospace Ground Test Facility Requirements*, recommended that budgets for facility upgrades, maintenance, and repair be increased and called attention to the upkeep of Tunnel 16T, AEDC's workhorse transonic testing facility, because of its importance to DoD programs (NRC, 1992).

Concerns over the viability of the major national low-speed and transonic test facilities led to a joint NASA-DoD initiative for a national wind tunnel complex. The project was originally envisioned as a two-tunnel complex that would cost about $2 billion. When the original cost estimates were shown to be too low, the plans were changed to one tunnel that could test for both speed ranges. The design for the tunnel was completed, but the program was dropped, and the tunnel was never built.

As the use of wind tunnel facilities has declined, interest in expanding the nation's aerodynamic test facilities has also declined. The focus today is on reducing operating and maintenance costs of existing facilities. Putting some facilities in a standby, or mothballed, status and decommissioning—even demolishing—others are being considered. However, a number of significant issues must be resolved before any of these steps are taken. For a better understanding of these issues, several members of the committee spent a day at AEDC. Although the following discussion is based on the visit to AEDC, it is also applicable to other government test complexes such as those operated by NASA.

SIMULATION AS A REPLACEMENT FOR PHYSICAL TESTING

An NRC study completed in 1983, entitled *The Influence of Computational Fluid Dynamics (CFD) on Experimental Aerospace Facilities*, warned of excessive dependence on computations as a substitute for physical testing. The study concluded (NRC, 1983) that

> the extensive application of CFD hinges upon two major considerations. First, the designer must have a high degree of confidence in the computational methods for aerodynamic design as compared to testing. Second, management from industry and government must have confidence that CFD is a more efficient developmental tool than extensive wind tunnel testing. For the next 15 years, CFD and ground test facilities will be used in a complementary mode with no appreciable reductions in testing anticipated.

This prediction has turned out to be correct. Although the performance of aerospace vehicles at design conditions can be reasonably simulated, performance near the operational limits (e.g., stalls, buffeting, drag rise) has to be verified by testing in ground-based facilities or in flight. Although the emphasis in testing has shifted, the DoD emphasis on meeting performance requirements has become more stringent, and as a result there has been no significant reduction in testing hours.

PAST USE OF TEST FACILITIES

The amount of testing at a particular facility is generally used as a basis for deciding whether to continue to maintain the facility in an operational status. However, past use is not a good predictor of future need. Very little testing has been done in some AEDC facilities in the past three years. However, the decision of whether to mothball or decommission these facilities should be based on the need for their capabilities in the future. The Air Force should include an estimate of the test facilities in defining the development of future systems (Heil, 2000).

SUPPORT FOR COMMONALITY

Large test complexes with a variety of individual facilities often share common support assets. A notable example is the common drive system at AEDC shared by the 16-foot transonic and the 16-foot supersonic tunnels. Other support, such as technician staff, instrumentation facilities, and model shops, is often shared among wind tunnels. Therefore, a decision to close only one test facility at a complex may not lead to significant total savings. Commonality should always be considered in the decision to deactivate a facility (Heil, 2000). Since industry is a user of these test facilities, it should also be a participant in the decision process.

RETENTION OF CRITICAL SKILLS

Most wind tunnel testing is done under considerable time pressure. It is not uncommon for the design process of specific details of an air vehicle to be suspended until wind tunnel test results are available. Modifications and additions to basic test facilities, as well as changes to operating modes, are not always recorded in any formal way; they may exist only in the memories of the operating personnel. In this sense, test facilities are similar to manufacturing facilities in which critical so-called black-book knowledge is vital to efficient operations. Therefore, decisions to deactivate a facility should take into consideration the effects of losing the personnel who know how to reactivate and operate the facility (Heil, 2000).

MODERNIZING AND UPDATING

The decline in test program activity in the United States has diminished incentives to modernize and improve test facilities. Most test centers have long lists of proposals for increasing the efficiency, lowering the cost, and expanding the envelope for testing and research. Because of budget limitations, most of these proposals have not been funded and are in an "on-hold" status.

For example, AEDC has a proposal to modify its Aerodynamic and Propulsion Test Unit cell to permit the testing of tactical missile systems and on-demand launch and recovery systems for low-cost access to space. "The Propulsion Wind Tunnel Cycle Time Reduction" proposal projects an increase of 25 testing days per year at a cost of $50 million. The modification would reduce the average time for installation or removal of a model from more than eight hours to four hours (AEDC, 2000).

These unfunded proposals have important implications for decisions to close or mothball facilities. AEDC has more than 60 defined, unfunded tasks. If the number of facilities is reduced, the effectiveness of the remaining facilities should be increased to offset the loss of test capacity (AEDC, 2000).

Before active test facilities are shut down, comparable lists of all government facilities should be reviewed to ensure that appropriate improvement and modification tasks of the remaining facilities are undertaken so as to guarantee maximum overall effectiveness.

Both NASA and AEDC are trying to reduce the operations and maintenance costs of existing facilities. Currently, NASA is changing to a full-cost accounting system. The cost at the NASA Ames Research Center, for example, would increase from $1,550 per hour plus the cost of power to $7,200 per hour plus the cost of power (NASA, 1999). AEDC is a government-owned, contractor-operated (GOCO) facility with cost procedure limitations that differ from NASA's (Heil, 2000). Therefore, the cost of testing can differ greatly, and the facility with lower user costs will probably be used more, even if it has lesser capabilities than the more expensive facility.

Because the use of a facility is uneven, with considerable periods of inactivity, reducing costs can have unanticipated effects. Direct cost accounting has certain advantages from the standpoint of financial management; the long-term effect on a facility that is not used extensively is to increase user costs. If an organization cannot afford the increased costs, it will simply stop using the facility. As a result, costs will go up for the remaining users, some of whom will then stop using the facility, and so on. If the facility is scarcely used, a decision may be made to close it altogether, even though the facility may be critical for testing future advanced systems. In the long term, new systems could entail higher system costs and perform less well than if they had been improved through testing in the facility. In conclusion, although they are expensive to maintain and operate and require continual investment to keep them efficient and up to date, major technical facilities, such as wind tunnels, will be needed in the future. Planning for future technical resources must include maintaining these facilities. Fiscal pressures on the testing agencies are intensified by the continued decline in the use of aeropropulsion test and wind tunnel facilities. These pressures could be alleviated somewhat if DoD and NASA agreed on which facilities will be needed. These facilities could then be properly maintained and upgraded, and excess facilities could be mothballed or closed. The cost of using comparable facilities should be as close to uniform as possible so that the choice of facility is based on availability of the required test capability rather than on price competition.

Facilities used solely for research could continue to operate at the discretion of the responsible agency. In that case, the Air Force would have to define its long-term system goals and translate them into specific requirements for test facilities.

RELATIONSHIP WITH INDUSTRY

As the commercial segment of the economy continues to increase, the burdens associated with defense contracts are becoming more difficult to justify and support. Because of their unwillingness to accept DoD acquisition rules, key commercial suppliers, such as Intel, no longer supply military-unique hardware as a matter of policy. Thus, the pool of available companies and technical talent from which DoD can draw is shrinking.

Federal Acquisition Regulations (FAR)—such as the requirement to apply government cost-accounting standards (CAS), the Truth in Negotiations Act (TINA) (Public Law 10 USC section 2306a), and the False Claims Act (Public Law 31 USC sections 3729-3733)—are frequently cited as major obstacles to efficient, effective relationships between the government and the industrial sector. Although everyone would agree with the idealistic intent of TINA, many of the regulations simply generate paperwork and increase administrative costs. Some progress is being made. For example, FARs do not permit companies to recover the full cost of benefits and incentives, such as moving allowances and certain stock options that companies may find necessary to offer to recruit new people in today's marketplace (LMSS, 2000).

The DSB recently recommended making recruitment and retention bonuses for people with critical skills, such as software and avionics specialists, recoverable (DSB, 2000b). In October 2000, Undersecretary of Defense for Acquisition and Technology Jacques Gansler announced that this policy change would be implemented. As another example, in 2000, Congress passed an initial reform of the government-unique CAS; the threshold on the value of contracts governed by the CAS was raised significantly (Douglass, 2000b).

In FAR Section 845, "Other Transactions Authority," Congress authorized the Defense Advanced Research Projects Agency (DARPA) to enter into agreements with private industry without adherence to the FAR. The intent was to allow DARPA to experiment in its contracts for R&D. Under Section 845, government organizations or agencies can establish relationships with industry much more akin to commercial relationships. Congress has since authorized the extension of Section 845 authority to the military services. In a report issued in 1999, however, the DSB concluded that application by DOD was still quite limited (DSB, 1999). DoD could use the opportunity to explore commercial-type arrangements with industry to evaluate a variety of alternative arrangements. Lessons learned could be captured and used to identify, evaluate, and provide support for meaningful refinements to FAR.

6

Conclusions and Recommendations

THE BASIC CONCLUSION

The technical resources problem is not separable from the Air Force's other duties. It affects and is affected by Air Force decisions about current and future missions and needs and what the Air Force develops, buys, tests, and uses in training, which in turn leads to what the future of the Air Force is to be. If the technical resources required are not considered when and where these fundamental matters are discussed and decided, their effects will not be properly taken into consideration. The Air Force pays close attention to those matters it holds most important. For many years, however, it enjoyed high-quality technical resources without paying specific attention to them, but times are changing. In the opinion of the committee, the technical resources will not continue to be of high quality without this specific attention.

In the committee's opinion, this problem is best dealt with by the Chief of Staff of the Air Force saying he wants attention paid to technical quality and quantity and then appointing a deputy for this purpose, preferably as a sole responsibility.

Further, although S&T is important, in fact very important, it is only a part of what concerns the committee. The committee believes it would be desirable for a deputy chief of staff (DCS) to have responsibility for S&T as a part of his larger job. Alternatively, someone else could have direct S&T responsibility, as is true for most of the other elements of the DCS's portfolio. This DCS, in the committee's view, would have oversight of all Air Force technical resources and how they fit into what the Air Force is doing and plans to do.

The basic conclusion of this report is that if the Air Force is concerned about the future of its technical resources (and it is), it must give the problem the kind of continuing attention that it gives to other serious matters. This attention includes the following:

- Raising the level of attention by establishing a deputy chief of staff who is also a member of the Air Force Council to oversee the Air Force's technical resources;
- Creating an ongoing assessment of Air Force technical resources and planning what to do to ensure their quality;
- Paying attention to all elements of the technical spectrum, including S&T, which is necessary although not sufficient, as well as R&D, design, and production;
- Considering separately the health of industry, in-house labs, the efficacy of program management, and the health of universities—and doing what is necessary to ensure their continued health in supporting the Air Force; and
- Making sure the technical community fully understands the Air Force's need for and commitment to high-quality, leading-edge technology and the role of Air Force technology and acquisition people.

SPECIFIC CONCLUSIONS AND RECOMMENDATIONS

The character of potential adversaries has changed, as has the environment of the defense aerospace industry. For the Air Force to maintain its technological lead, it too must change if it is to maintain the quality of its technical resources. The major areas of change should be in the management of scientific and technical resources, maintenance of the technical work force, the Air Force's relationship with industry and academia, and related government industrial policies. The committee's major conclusions and recommendations in these areas are presented below.

Scientific and Technical Resources

A shrinking budget and some insistence on maintaining business as usual have led to great instabilities. The Air Force has more programs, particularly new procurement programs, than it can support with its present and foreseeable acquisition budgets. As a result, funds are constantly being shifted and reprogrammed to pay pressing bills. This has created unnecessary, debilitating upheavals for both personnel and programs and may prejudice the Air Force's long-term technology initiatives.

The committee recognizes that the Air Force major program budget is heavily affected by the political process. Nevertheless, the Air Force should decide upon, and then protect, the portion of the budget allocated for future technologies, which will determine the quality of its future warfighting capability.

Conclusion 1. In the process of refocusing its priorities and as a result of reorganizations predicated by the Goldwater-Nichols Act, the Air Force eliminated the position of Deputy Chief of Staff for Research and Development and thus lost a strong advocate for science and technology. In addition, the Air Force Systems Command was combined with the Air Force Logistics Command to form the Air Force Materiel Command (AFMC). Although this consolidation has streamlined AFMC's processes for development, acquisition, and support of Air Force systems, it has also reduced the emphasis on technology in general and S&T in particular. Currently, the highest S&T-dedicated position in the Air Force is the two-star Air Force Research Laboratory (AFRL) commander position at Wright-Patterson Air Force Base (AFB) near Dayton, Ohio, which is several levels below the Air Force Council. The AFRL commander reports directly to a general (four-star), the commander of AFMC, of which AFRL is a part. AFMC headquarters is also located at Wright-Patterson. The AFMC commander's responsibilities are very broad, including the programs at four product centers, five air logistics centers, three test centers, and two major specialized centers, in addition to AFRL. The AFMC commander has too many other important and demanding responsibilities to focus on S&T and without an S&T advocate of sufficient stature and authority at the Air Staff level where budget decisions are made, support for S&T has declined substantially. Reinstating a senior voice for technology in shaping the future capability of the Air Force would help ensure that decisions affecting technical resources including S&T expenditures are fully considered by Air Force decision makers.

Goldwater-Nichols defines the relationship between the Secretary's Office and the Chief's Office on technical matters. The Secretary's responsibilities are clear, but this should not mean the military does not have very strong concerns about and influence on the technical resources of the Air Force and does not have the need for oversight. The committee believes that the Air Force, both civilian and military, must pay more focused attention to its technical resources if it is to continue to get the best weaponry.

If Goldwater-Nichols or other constraints make a DCS position unworkable, the role and responsibility recommended could be assigned in other ways. The committee understands, for example, that the Navy has established a position for a two-star Director of Test & Evaluation and Technology Requirements (N091) who reports directly to the Chief of Naval Operations (CNO) and has somewhat similar functions. He serves as the principal interface between the CNO and the Assistant Secretary of the Navy for Research, Development, and Acquisition on RDT&E; Resource Sponsor for Navy S&T (6.1, 6.2, 6.3A) investments; Resource Sponsor for RDT&E field activities; and Appropriations Sponsor for CNO RDT&E, Navy (RDT&E,N) funding. In any event, the committee believes that high-level attention is needed to get the best result.

Recommendation 1. The Air Force should establish a deputy chief of staff, who is also a member of the Air Force Council, with primary responsibility for oversight of all Air Force scientific and technical resources. Among his duties, which should include all Air Force technical activities from concept development through completion of engineering and manufacturing development (EMD) phases, this officer should be the advocate for funding science and technology requirements and for modifying and tracking the implementation of S&T requirements to minimize instabilities in S&T and R&D funding (including new production processes), to ensure that adequate funding is budgeted annually, and to resist attempts to raid S&T or R&D funds to meet short-term budget shortfalls in other areas.

The committee believes that whoever is assigned the responsibilities for oversight of Air Force technical resources should be able to act as a high-level advocate for the technical resources within the Air Staff and the department. This individual should be someone with an extensive scientific or technical education and background, yet who also has experience in the operational commands and can appreciate the critical needs of both sides of the house—warfighting and technical. The new DCS would maintain a awareness of the status of all aspects of the Air Force's technical resources and would track the effects of current and proposed policies—concerning personnel, facilities, the Air Force Institute of Technology (AFIT), education, and research activities—on the technical capabilities base. This person would then serve as an advocate on the Air Staff to ensure that the needs of the technical capabilities base at least get a fair and accurate hearing in the policy decisions of the Air Force.

The Air Force Technical Work Force

Despite the recent difficulties of attracting qualified people to government defense work, the committee believes the Air Force has marginally enough scientific, technical, and engineering personnel to carry out its current programs and, with the appropriate effort, should be able to attract enough people in the near future to develop and build the systems that are now planned. In fact, the Air Force, even with reduced S&T funding, has the resources to pursue many important programs on the leading edge of technology, providing it sets the right priorities and executes them efficiently. The problem is how to attract new talent in the face of growing commercial competition. This is a new challenge for the Air Force, and meeting it will require a new way of thinking. The committee strongly supports technical education for Air Force personnel at both the Air Force Institute of Technology and civilian universities as a source of technically educated officers for S&T and acquisition force positions.

Conclusion 2. Technical personnel—having the types and numbers that are essential for maintaining current and future Air Force technological superiority and for ensuring technical innovations in science, technology, research, and development to support current, emerging, and future capabilities—are just as important to the successful accomplishment of the Air Force mission as the operational and support elements.

Recommendation 2. The Air Force should assess the quality and quantity of its technical personnel regularly, taking into consideration its future missions and needs, just as it currently assesses the quality and quantity of its flight personnel and other vital resources. The Air Force should use these assessments to define the types and numbers of technical personnel necessary to maintain current and future Air Force technological superiority and ensure technical innovations in science, technology, research, and development. The results of the assessment should be used as a basis for making policy changes (if necessary) to protect investments that support technical personnel.

Ongoing assessments should include the following estimates:

- The number of uniformed and civil service technical personnel necessary now and in the future, including skills, technical specialties, and years of experience;
- The organizational base and operating requirements to support the Air Force's long-term technology needs, including size, skills, and responsibilities for universities and in-house laboratories performing scientific and technical research, industrial contractors that the Air Force considers essential to maintaining a competitive supplier base, organizations the Air Force intends to sustain as sole suppliers in particular areas, and Federally Funded Research and Development Centers (FFRDCs); and
- The need for research, development, test, and evaluation (RDT&E) facilities, such as wind tunnel facilities and test ranges, for transitioning technology capabilities and meeting research requirements. This is included because high-quality technical people need and will insist on high-quality facilities. These needs should be closely coordinated with the other services and with the National Aeronautics and Space Administration (NASA).

This assessment should be the responsibility of the deputy chief of staff described above. Finally, that officer should present this assessment of Air Force technical resources, both current and future, to the senior Air Force leadership annually during the periodic meeting at Corona.

Conclusion 3. With a nearly 35 percent drop in the Air Force S&T budget over the last 15 years, the Air Force cannot sustain its technical viability over the long term. Although total S&T funding for DoD as a whole is on the order of 4 percent of the DoD budget, the Air Force has selectively reduced the percentage it allocates for this purpose compared to the other services. The Air Force reductions in S&T funding also come at a time when other sources of federal S&T funding which support the Air Force, such as NASA and Federal Aviation Administration (FAA) research, have declined significantly. S&T funding is the "seed corn" for the technical superiority of the Air Force. Therefore, the level of S&T funding must be maintained at an adequate level.

Recommendation 3. The Air Force should balance current expenditures and investments in future technologies and insulate those budgets from the vagaries of near-term fiscal pressures. Vital science and technology resources should be organized, protected, and nurtured just as carefully as critical operational resources.

Relationship with Industry

Few, if any, perfectly free markets exist anywhere with many suppliers and many buyers, perfect information, and no applied restraints. The DoD as a monopsony, or single buyer, for the defense industry cannot be said to operate in anything like a free market. This, however, does not mean that there is no competition, just that the competitions are established and controlled by DoD. The DoD has widely varying relationships with its suppliers ranging from open competitions to what are essentially permanent single sources and all other combinations in between. Since DoD sets the rules, it is responsible for the effects of those rules on its supplier base whether it recognizes this explicitly or not. The committee understands that this is well recognized within DoD but that there seems to be no established mechanism for determining and including the effects on the Air Force's technical resources when decisions are being made about the competitive conditions for individual programs or for the Air Force program as a whole.

In a free market, the responsibility for maintaining a high-quality technical staff can and should be left to suppliers. However, the aerospace defense industry is not a free market. The Air Force and DoD as a whole are a monopsony customer for much of the aerospace industry and therefore have significant influence over the supplier base and infrastructure. Despite changes in DoD policy requiring that all of the services increase their use of commercial off-the-shelf products and modify their procedures to accommodate commercial practices, the Air Force still depends on the aerospace defense industry for high-quality, Air Force-unique products.

Through its procurement practices, the Air Force can affect, even determine, the long-term viability of these suppliers.

Conclusion 4. In the current environment, the Air Force cannot continue to think of the defense aerospace industry as a competitive marketplace. The way the Air Force uses its resources, by design or otherwise, has a major impact on the viability of the defense aerospace infrastructure that supports the national security interests of the Air Force. Maintaining the traditional arms-length relationship with industry must yield to establishing long-term partnerships with responsibilities on both sides. Partnership includes the concepts of mutually supportive and collaborative relationships and two-way exchange in which the parties depend on each other. Industrial organizations must be responsible for maintaining their own capabilities, but the Air Force must be responsible for providing conditions and incentives under which these organizations can remain strong and effective and continue to enhance their technical capabilities.

Recommendation 4. Air Force management should take into consideration the effects of its budget and management policies on industry. The Air Force should establish partnerships with defense industries that will encourage industry's continued effectiveness and enhance its technical capabilities.

A number of studies analyzing the defense industrial environment were reviewed by the committee. These studies largely agree that the needs of the industrial base should be taken into account in the acquisition process, that industry metrics must be better understood, and that the export control process must be streamlined. The committee's investigations substantiated the results of these studies and the validity of their recommendations.

Although government oversight of the defense industry is necessary to ensure that taxpayer money is spent wisely and appropriately, current government regulations are focused excessively on detailed cost accounting and adherence to complex processes. As a result of these regulations, efficiency has been undermined, procurement costs have risen substantially, and the attractiveness to technical people of working in defense has been reduced. Unless high-quality people are treated as trustworthy and reliable, they will go elsewhere. Some leading technology corporations, such as AT&T, Unisys, Honeywell, Westinghouse, Ford, GM, and others,[1] now are reluctant to compete for new U.S. government contracts, preferring to offer standard products on a fixed-price basis. These corporations have divested themselves of divisions and subsidiaries that made defense products (examples are the divisions of Westinghouse and Hughes (GM) that made airborne fire-control radars). These divisions and subsidiaries were sold to companies that had similar or related products, and the capabilities of the acquiring and acquired industrial units, although downsized and merged, remain available to the DoD. The sell-offs appear to have been driven by two main factors. First, they were encouraged to merge or sell and, at least in the early part of the decade, almost coerced to do so by DoD in recognition of its reduced weapon acquisition programs after the end of the Cold War, under the heading of "consolidation." Second, the prevailing equity market mantra generally rewarded corporations for sticking to their core businesses and for getting rid of fringe activities, particularly if they had lower profitability and/or lower growth potential than the corporation's mainstream business.

The Air Force would be helped a great deal if it could work more easily and efficiently with commercial organizations in buying commercial products and taking advantage of commercial technology and processes. New initiatives for modifying the Federal Acquisition Regulations (FAR) should enable a shift toward more commercial-like procurement relationships.

Conclusion 5. Current government reforms are a step in the right direction toward dramatically reducing acquisition cycle times and will certainly improve the efficiency of the defense industry. They will also allow greater flexibility in program design and result in processes that are more in tune with industry. Reform should be considered a long-term undertaking and should continue to be visibly supported by Air Force leadership. The issue of whether reducing cycle time will increase near-term budgets depends on whether one is attempting to squeeze all of the existing programs into a shorter time or whether one is trying to do as many as can reasonably be done within the available funds. Programs that cannot be fitted in will be delayed. Over a long period, more programs would get done within the same total funds, most of them sooner than they could be now because of the shortened cycle time and greater efficiency.

Recommendation 5. The Air Force, and the U.S. Department of Defense as a whole, should continue to provide strong leadership for initiatives to reduce acquisition cycle times as a means for furthering DoD goals. They should also continue to work toward reforming policies and regulations for acquisition processes, thereby reducing the burden on industry of working with the government.

Reforming Policy and Regulations

In addition to interesting work, quality people need quality working conditions. A healthy organization, whether in industry or in government, must provide competitive salaries, benefits, and opportunities for growth and advancement. Civil service regulations, which apply to the government service white-collar administrative force, the blue-collar work force, and the technical work force, use standardized processes that are poorly suited to meet the demands of gov-

[1]GE and IBM are still performing substantial ongoing work for DoD. For example, GE still has a military jet engine division.

ernment laboratories performing leading-edge research. Inflexible civil service regulations, hiring practices, employment conditions, and salary structures have resulted in a serious decline in government technical talent, and more and more bench-scale technical work is being contracted out to industry and university laboratories.

This adverse trend has exacerbated the problem of the viability and sustainability of government laboratories. Leaders of government technical organizations have been working hard to overcome these difficulties but have not received adequate policy and regulatory support. In the committee's opinion, the supply of good technical people the Air Force needs cannot be maintained under existing civil service rules and restrictions. Therefore, the Air Force must actively pursue civil service reform for scientific and technical personnel. A Defense Science Board (DSB) Task Force addressing the ability of DoD to attract and retain critical personnel recommended the transfer of authority for the DoD civilian work force from the Office of Personnel Management (OPM) to the Secretary of Defense. The committee did not endeavor to evaluate this proposal but believes that this concept merits serious consideration.

Conclusion 6. The Air Force is facing serious competition for high-quality technical people. Although the Air Force can offer interesting and important work for in-house government employees and has a substantial budget, it has had trouble attracting people either directly out of universities or from the commercial sector. Many of the difficulties are related to civil service rules and restrictions that do not allow the salary flexibility and rapid decisions necessary to hire talented available applicants quickly. In order to meet the need to compete more effectively than it does now, the Air Force must provide a quality workplace.

Recommendation 6. The Air Force should join the Office of the Secretary of Defense, the other services, and other federal agencies in ongoing attempts to reform the civil service rules for scientific and technical personnel.

Universities are the trainers and motivators, the breeding ground, of the future defense work force. Air Force funding for university research is essential for maintaining this connection. The Air Force should cultivate and establish long-term partnerships with universities.

Conclusion 7. The universities, faculties, and students whose research is supported by U.S. government funds are an indispensable base for motivating young, well-educated individuals to pursue careers in defense technology and Air Force laboratories. The relationship between universities and the Air Force is symbiotic. Universities need S&T funds from the Air Force, and the Air Force needs both the results of S&T and new technical personnel. If the Air Force wants a healthy relationship it must cultivate and establish long-term partnerships with universities.

Recommendation 7. The Air Force should establish long-term, stable partnerships with its supporting universities and their faculty members. The Air Force should decide how much to invest for the future through S&T funds to universities and then protect that investment. The Air Force should also recognize the financial problems facing universities and make sure that contractual and financial arrangements are consistent with the continued health of these important institutions and their ability and willingness to continue to support the Air Force.

The Air Force must do more to attract and retain the highest-quality scientific and technical people. As a matter of first principle, highly trained technical people want interesting and important work, which the Air Force has in plentiful supply. Despite reduced R&D budgets, the Air Force continues to devote major funds to solving leading-edge technical challenges. However, the impression in industry and the technical community is that the Air Force is short of money and is concentrating its efforts on operational requirements, modifications, and upgrades at the expense of new technology. Although the Air Force's emphasis has recently been focused on operational requirements for multiple force deployments, substantial funds are still being invested in interesting and challenging S&T projects. The Air Force should develop a strong positive message about its technical S&T program.

Conclusion 8. Air Force technical programs and opportunities are challenging and exciting. However, the Air Force has not communicated that excitement to the technical community. The Air Force must overcome the perception that opportunities in defense research are limited and that defense is not as important as it was during the Cold War.

Recommendation 8. The Air Force should communicate a strong, positive message describing its technical plans and opportunities and ensure that this message is broadly distributed to students, faculties, industry, and the general technical community.

References

AEDC (Arnold Engineering Development Center). 2000. Unfunded requirements list. Tullahoma, Tenn.: Arnold Engineering Development Center.

AIA (Aerospace Industries Association). 2000. Aerospace Facts and Figures 1999/2000. Washington, D.C.: Aerospace Industries Association.

AIA. 2001a. Aerospace Facts and Figures 2000/2001. Washington, D.C.: Aerospace Industries Association.

AIA. 2001b. Total U.S. aerospace sales for 1985 and 1997. Data provided to the Committee on the Future of the U.S. Aerospace Infrastructure and Aerospace Engineering Disciplines to Meet the Needs of the Air Force and the Department of Defense, July 2, 2001.

Anderson, C.M. 2000. Air Force X vehicle technology. Briefing by C.M. Anderson, Chair, Space Platforms Defense Technology Area Panel, to AIAA X Vehicles Symposium, Capitol Hill Club, Washington, D.C., June 16.

Augustine, N.R. 1996. From industry . . . Acquisition reform: Dream or mirage? Army RD&A Bulletin 5:20.

BLS (Bureau of Labor Statistics). 1999. Employment projections; employment, training, and earnings; education level search; percent change in total employment, 1998–2008. Available online at <http://stats.bls.gov/asp/oep/noeted/empeduc.asp> [April 24, 2001].

Boeing. 2000a. F/A-18E/F Super Hornet fact sheet. Available online at <http://www.boeing.com/defense-space/military/fa18ef/fa18ef.htm> [February 9, 2001].

Boeing. 2000b. Presentation to fact-finding subcommittee of the Committee on the Future of the U.S. Aerospace Infrastructure and Aerospace Engineering Disciplines to Meet the Needs of the Air Force and the Department of Defense, Boeing, St. Louis, Mo. April 6.

Boeing. 2001a. F-15 Eagle fact sheet. Available online at <http://www.boeing.com/defense-space/military/f15/f15.htm> [February 9, 2001].

Boeing. 2001b. C-17 Globemaster III fact sheet. Available online at <http://www.boeing.com/defense-space/military/c17/c17.htm> [February 9, 2001].

Boeing. 2001c. 767 Tanker-Transport fact sheet. Available online at <http://www.boeing.com/defense-space/military/767t-t/ > [February 9, 2001].

Capaccio, T. 2000. Pentagon enacting initiatives to boost defense industry. Defense Week, October 30:1.

Carlson, H.C. 2000. Air Force Research Laboratory technology vision. Briefing by H.C. Ruck, Chief Scientist, Air Force Office of Scientific Research, to the Committee on the Future of the U.S. Aerospace Infrastructure and Aerospace Engineering Disciplines to Meet the Needs of the Air Force and the Department of Defense, National Research Council, Washington, D.C., August 23.

CDM (Commission on Defense Management). 1986. A Formula for Action: A Report to the President on Defense Acquisition, April 1986. Commission on Defense Management. Ft. Belvoir, Va.: Defense Technical information Center.

CNN (Cable News Network). 2000. Aerospace industry eyes dot-com demise with glee. Available online at <http://www.cnn.com/2000/TECH/computing/12/15/brain.drain.reut/index.html> [April 30, 2001].

Creedon, J.F. 2000. NASA Langley Research Center. Briefing by J. Creedon, Director, NASA Langley Research Center, to the Committee on the Future of the U.S. Aerospace Infrastructure and Aerospace Engineering Disciplines to Meet the Needs of the Air Force and the Department of Defense, National Research Council, Washington, D.C., March 9.

CSAF (Chief Scientist of the Air Force). 1999. Science and Technology Workforce for the 21st Century. Washington, D.C.: Office of the Chief Scientist of the Air Force.

DARPA (Defense Advanced Research Projects Agency). 2000. Unmanned combat air vehicle (UCAV). Available online at <http://www.darpa.mil/tto/programs/ucav.html> [March 28, 2001].

David, L. 2000. X-37: An orbital test bed sprouts wings. Aerospace America 38(6):39–43.

DDR&E (Director, Defense Research & Engineering). 2001. S&T budget figures provided to the Committee on the Future of the U.S. Aerospace Infrastructure and Aerospace Engineering Disciplines to Meet the Needs of the Air Force and the Department of Defense, May 2.

Delaney, L.J. 1999. Leadership corner. Aerospace Acquisition 2000 2(3):2.

Douglass, J.W. 2000a. AIA. Briefing by J.W. Douglass, President, Aerospace Industries Association, to the Committee on the Future of the U.S. Aerospace Infrastructure and Aerospace Engineering Disciplines to Meet the Needs of the Air Force and the Department of Defense, National Research Council, Washington, D.C., January 27.

Douglass, J.W. 2000b. 106th Congress: Industry tallies major gains on aerospace issues. AIA Update 5(6):2.

DSB (Defense Science Board). 1999. Report of the Defense Science Board Task Force on Acquisition Reform Phase IV, Subpanel on Research and Development. Ft. Belvoir, Va.: Defense Technical Information Center.

DSB. 2000a. Report of the Defense Science Board Task Force on Human Resources Strategy. Washington, D.C.: Office of the Under Secretary of Defense for Acquisition, Technology, and Logistics.

DSB. 2000b. Report of the Defense Science Board Task Force on Preserving a Healthy and Competitive U.S. Defense Industry to Ensure Our Future National Security. Washington, D.C.: Defense Science Board.

Eash, J. 1998. Revolution in business affairs. Presentation by J. Eash, Deputy Under Secretary of Defense (Advanced Technology), to the Eighth PEO/SYSCOM Commanders Conference, Fort Belvoir, Va., October 19-20.

Erwin, S.I. 2001. Next-generation air-mobility concepts focus on efficiency. National Defense LXXXV(567):16–17.

Etter, D.M. 2000. S&T/engineering challenges and insights. Briefing by D.M. Etter, Deputy Under Secretary of Defense (Science and Technology), to Committee on the Future of the U.S. Aerospace Infrastructure and Aerospace Engineering Disciplines to Meet the Needs of the Air Force and the Department of Defense, National Research Council, Washington, D.C., April 4.

EWC (Engineering Workforce Commission). 2001. Engineering and technology degrees, 2000. Washington, D.C.: American Association of Engineering Societies.

GAO (General Accounting Office). 1997. Trends in DOD Spending, Industrial Production, and Competition. Washington, D.C.: General Accounting Office.

Harbison, J.R., T.S. Moorman, M.W. Jones, and J. Kim. 2000. Booz-Allen & Hamilton Viewpoint: U.S. defense industry under siege—An agenda for change. Available online at <http://www.bah.com/viewpoints/defenseindustry.html > [December 11, 2000].

Hastings, D. 2000. S&T/engineering personnel challenges and insights. Briefing by D. Hastings, Co-director, MIT Technology and Policy Program, to the Committee on the Future of the U.S. Aerospace Infrastructure and Aerospace Engineering Disciplines to Meet the Needs of the Air Force and the Department of Defense, National Research Council, Washington, D.C., April 4.

Heeter, E., and S. Kosiak. 2000. FY01 defense appropriations conference bill, July 28. Available online at <http://www.csbaonline.com/> [May 16, 2001].

Heil, M. 2000. Presentation by M. Heil, Commander, Arnold Engineering Development Center, to fact-finding subcommittee of the Committee on the Future of the U.S. Aerospace Infrastructure and Aerospace Engineering Disciplines to Meet the Needs of the Air Force and the Department of Defense, Arnold Engineering Development Center, Tullahoma, Tenn. October 12.

Johnson, C.J. 1995. Martin Marietta's CEO Norm Augustine speaks to Program Manager. Program Manager 24(2):2–6.

JSF PO (Joint Strike Fighter Program Office). 2001a. JSF Program white paper. Available online at <http://www.jast.mil/html/jsfwhitepaper.htm> [February 9, 2001].

JSF PO. 2001b. The affordable solution—JSF. Available online at <http://www.jast.mil/html/progbriefs.htm> [February 9, 2001].

Kennedy, M., and M. Lorell. 2000. Project Air Force. Briefing by M. Kennedy and M. Lorell, RAND Corporation, to the Committee on the Future of the U.S. Aerospace Infrastructure and Aerospace Engineering Disciplines to Meet the Needs of the Air Force and the Department of Defense, National Research Council, Washington, D.C., January 27.

LAI (Lean Aerospace Initiative). 1996a. 1996 LAI Government PM Survey. Cambridge, Mass.: Massachusetts Institute of Technology Lean Aerospace Initiative.

LAI. 1996b. 1996 LAI Contractor PM Survey. Cambridge, Mass.: Massachusetts Institute of Technology Lean Aerospace Initiative.

LM (Lockheed Martin Company). 2000. F-16 Fighting Falcon fact sheet. Available online at <http://www.lockheedmartin.com/factsheets/product16.html> [February 9, 2001].

LM. 2001. F-22 program overview. Available online at <http://www.lmasc.lmco.com/f22/mgoview.html> [February 9, 2001].

LMAC (Lockheed Martin Aeronautics Company). 2000. Presentation to fact-finding subcommittee of the Committee on the Future of the U.S. Aerospace Infrastructure and Aerospace Engineering Disciplines to Meet the Needs of the Air Force and the Department of Defense, Lockheed Martin Aeronautics Company, Fort Worth, Tex., February 25.

LMSS (Lockheed Martin Space Systems). 2000. Presentation to fact-finding subcommittee of the Committee on the Future of the U.S. Aerospace Infrastructure and Aerospace Engineering Disciplines to Meet the Needs of the Air Force and the Department of Defense, Lockheed Martin Space Systems, Littleton, Colo., April 18.

London, J.R. 2000. Pathfinder Program. Briefing by J.R. London, Manager, Technology Evaluation Department, to AIAA X Vehicles Symposium, Washington, D.C., June 16.

Lorell, M.A. 1995. Bomber R&D Since 1945. Santa Monica, Calif.: RAND Corporation.

Lorell, M.A., and H.P. Levaux. 1998. The Cutting Edge: A Half Century of U.S. Fighter Aircraft R&D. Santa Monica, Calif.: RAND Corporation.

Lorell, M., J. Lowell, M. Kennedy, and H.P. Levaux. 2000. Cheaper, Faster, Better? Commercial Approaches to Weapons Acquisition. Santa Monica, Calif.: RAND Corporation.

Lyles, L.L. 1999. Achieving cycle time reductions. Aerospace Acquisition 2000 2(3):1.

McNutt, R. 1998. Reducing DoD Product Development Time. Cambridge, Mass.: Massachusetts Institute of Technology Lean Aerospace Initiative.

Morrocco, J.D. 2001. U.K. antes up $2.8 billion for JSF. Aviation Week & Space Technology 154(4):33.

NASA (National Aeronautics and Space Administration). 1999. Report of the Rotorcraft Subcommittee. Unpublished report from meeting at NASA Ames Research Center, August 24-25, 1999. NASA Aero-Space Technology Advisory Committee. Washington, D.C.: National Aeronautics and Space Administration Headquarters, Code R.

NASA. 2000. Report on Project Management in NASA by the Mars Climate Orbiter Mishap Investigation Board, March 13, 2000. Washington, D.C.: National Aeronautics and Space Administration.

Northrop Grumman. 2000. Presentation to fact-finding subcommittee of the Committee on the Future of the U.S. Aerospace Infrastructure and Aerospace Engineering Disciplines to Meet the Needs of the Air Force and the Department of Defense, Northrop Grumman, Arlington, Va., August 23.

NRC (National Research Council). 1983. The Influence of Computational Fluid Dynamics on Experimental Aerospace Facilities. Washington, D.C.: National Academy Press.

NRC. 1988. Review of Aeronautical Wind Tunnel Facilities. Washington, D.C.: National Academy Press.

NRC. 1992. Future Aerospace Ground Test Facility Requirements for the Arnold Engineering Development Center. Washington, D.C.: National Academy Press.

NRC. 1999. Recent Trends in U.S. Aeronautics Research and Technology. Washington, D.C.: National Academy Press.

NRC. 2001. Aging Avionics in Military Aircraft. Washington, D.C.: National Academy Press.

NSF (National Science Foundation). 2000. Federal Funds for Research and Development: Fiscal Years 1998, 1999, 2000. Arlington, Va.: National Science Foundation.

NSF. 2001. Federal Funds for Research and Development: Fiscal Years 1999, 2000, 2001. Arlington, Va.: National Science Foundation.

OASD (PA) (Office of the Assistant Secretary of Defense (Public Affairs)). 2000. OASD (PA) Press release, December 26, 2000. Available online at <http://www.defenselink.mil/news/Dec2000/c12262000_ct766-00.html> [February 9, 2001].

OUSD (AT&L) (Office of the Under Secretary of Defense (Acquisitions, Logistics, and Technology)). 2000. The Road Ahead: Accelerating the Transformation of DoD Acquisition and Logistics Processes and Practices. Washington, D.C.: Office of the Under Secretary of Defense (Acquisitions, Logistics, and Technology).

Paul, R. 2000. Science and Technology Workforce for the 21st Century (STW-21). Briefing by R. Paul, Commander, Air Force Research Laboratory, to the Committee on the Future of the U.S. Aerospace Infrastructure and Aerospace Engineering Disciplines to Meet the Needs of the Air Force and the Department of Defense, National Research Council, Washington, D.C., March 10.

Reinsh, B. 2000. Testimony from Bill Reinsh, Department of Commerce, before the Senate Foreign Relations Committee, June 7.

Ruck, H. 2000. AFRL in the FYDP. Briefing by H. Ruck, Associate Director, Directorate of Plans and Programs, Air Force Research Laboratory, to the Committee on the Future of the U.S. Aerospace Infrastructure and Aerospace Engineering Disciplines to Meet the Needs of the Air Force and the Department of Defense, National Research Council, Washington, D.C., August 23.

Seffers, G.I. 2000. Finding a way to fix the brain drain. Federal Computer Week, July 31. Available online at <http://www.fcw.com/fcw/articles/2000/0731/mgt-drain-07-31-00.asp> [October 31, 2000].

Shelton, G. 2000. Attracting talent in the defense industry. Briefing by G. Shelton, Vice President, Raytheon Missile Systems, to the Committee on the Future of the U.S. Aerospace Infrastructure and Aerospace Engineering Disciplines to Meet the Needs of the Air Force and the Department of Defense, National Research Council, Washington, D.C., April 4.

Smith, P.G., and D.G. Reinertsen. 1995. Developing Products in Half the Time. New York: Van Nostrand Reinhold.

Stewart, J.D. 2000. AFMC workforce reshaping. Briefing by J.D. Stewart, Executive Director, Air Force Materiel Command, to the Committee on the Future of the U.S. Aerospace Infrastructure and Aerospace Engineering Disciplines to Meet the Needs of the Air Force and the Department of Defense, National Research Council, Washington, D.C., April 5.

Stone, E. 2000. Presentation by E. Stone, Director, Jet Propulsion Laboratory, to fact-finding subcommittee of the Committee on the Future of the U.S. Aerospace Infrastructure and Aerospace Engineering Disciplines to Meet the Needs of the Air Force and the Department of Defense, Jet Propulsion Laboratory, Pasadena, Calif., February 23.

Thompson, L.B. 2000. Into the night: The fin-de-siecle defense industry. Briefing by L.B. Thompson, chief operating officer, Lexington Institute, to Lockheed Martin, Arlington, Va., January 18.

USAF (United States Air Force). 2000a. CSAF displays confidence in, commitment for V-22, press release, June 28, 2000. Available online at <http://www.af.mil/news/Jun2000/n20000628_000986.html> [February 9, 2001].

USAF. 2000b. Biographies of senior Air Force acquisition personnel. Available online at <http://www.af.mil/news/biographies/> [February 21, 2001].

USD(C) (Under Secretary of Defense (Comptroller)). 2000a. RDT&E Programs (R-1), February 2000. Washington, D.C.: Under Secretary of Defense (Comptroller).

USD(C). 2000b. National Defense Budget Estimates for FY 2001, March 2000. Washington, D.C.: Under Secretary of Defense (Comptroller).

USN (United States Navy). 1999. V-22A Osprey. Available online at <http://www.chinfo.navy.mil/navpalib/factfile/aircraft/air-v22a.html> [February 9, 2001].

Velocci, A.L. 2001. Defense acquisition reform a hot issue, despite past failures. Aviation Week & Space Technology 154(9):65–66.

Venneri, S. 2000. Presentation by S. Venneri, Associate Administrator, NASA Office of Aerospace Technology, to fact-finding subcommittee of the Committee on the Future of the U.S. Aerospace Infrastructure and Aerospace Engineering Disciplines to Meet the Needs of the Air Force and the Department of Defense, National Aeronautics and Space Administration, Washington, D.C., March 8.

Wall, R. 2000. USAF readies strategy for increased F-22 buy. Aviation Week & Space technology 153(21):30–31.

Womack, J.P., and D.T. Jones. 1996. Lean Thinking: Banish Waste and Create Wealth in Your Corporation. New York: Simon & Schuster.

Appendixes

Appendix A

Meetings and Activities

FIRST MEETING

December 9-10, 1999
Sheraton National Hotel
Columbia Pike and Washington Blvd.
Arlington, Va.

Origin of Study and Air Force Expectations
William Berry
Air Force Research Laboratory

SECOND MEETING

January 27-28, 2000
Holiday Inn Georgetown
Kaleidoscope Room
2101 Wisconsin Avenue, N.W.
Washington, D.C.

Project Air Force
Mike Kennedy
RAND Corporation

Recent Trends in U.S. Aerospace Research and Technology
George Levin
National Research Council's Aeronautics and Space Engineering Board

Development and Current Status of the Aerospace Industry
John Douglass
Aerospace Industries Association (AIA)

Assessing Industrial Capabilities
Martin Meth
Industrial Capabilities and Assessments (OUSD (A&T)/IA)

Department of Defense (DoD) Science and Technology (S&T)
James Garcia
National Research Council's Committee on Review of the Department of Defense Air and Space Systems Science and Technology

Engineering Education Trends in the United States
Eli Reshotko
Case Western Reserve University

Pre-World War II R&D Funding Trends Subcommittee Discussion
Thomas Perdue
The Signature Group

THIRD MEETING

March 9-10, 2000
Holiday Inn Georgetown
Kaleidoscope Room
2101 Wisconsin Avenue, N.W.
Washington, D.C.

National Aeronautics and Space Administration (NASA) Engineering Disciplines Challenges Presentation
Jeremiah Creedon
NASA Langley Research Center

Lockheed Martin Fact-finding Meeting
Gordon England
Lockheed Martin Aeronautics Corporation (LMAC)

Lockheed Martin Fact-finding Meeting
Terry Alfriend
Texas A&M University

Historic Budget Data Follow-up
Thomas Perdue
The Signature Group

Defense Advanced Research Projects Agency (DARPA) Fact-finding Meeting
Thomas Perdue
The Signature Group

Jet Propulsion Laboratory Fact-finding Meeting
George Paulikas
The Aerospace Corporation

Discussions with Dr. Venneri, NASA
Raymond Colladay
RC Space Enterprises, Inc.

Academia Follow-up
Win Phillips
University of Florida

Academia Follow-up
Eli Reshotko
Case Western Reserve University

Academia Follow-up
Terry Alfriend
Texas A&M University

Science and Technology Work Force for the 21st Century Initiative
MG Richard R. Paul
Air Force Research Laboratory

FOURTH MEETING

April 4-5, 2000
Holiday Inn Georgetown
Kaleidoscope Room
2101 Wisconsin Avenue, N.W.
Washington, D.C.

Science and Technology/Engineering Personnel Challenges and Insights
Daniel Hastings
Massachusetts Institute of Technology

Air Force Materiel Command (AFMC) Workforce Shaping Initiative
J. Daniel Stewart
Air Force Materiel Command

Science and Technology and Engineering Challenges and Insights
Delores Etter
Undersecretary of Defense (Science and Technology)

Historic Budget Data Follow-up
James Myska, Research Associate
Committee on the Future of the U.S. Aerospace Infrastructure and Aerospace Engineering Disciplines to Meet the Needs of the Air Force and the Department of Defense

Engineering Hiring and Retention Challenges
Gregory Shelton
Raytheon Missile Systems

Attracting New Engineers to Aerospace Industry
Bob Ormsby
Lockheed Aeronautical Group Systems (retired)

Ten Things Air Force-Department of Defense (DoD) Can Do
Win Phillips
University of Florida

Space System Fact-Finding
Raymond Colladay
RC Space Enterprises, Inc.

FIFTH MEETING

June 23, 2000
National Research Council
Green Building
2001 Wisconsin Avenue, N.W.
Washington, D.C.

Air Force Scientist and Engineering Work Force
Brendan Godfrey
Brooks Air Force Base

Scientist and Engineering Hiring, Physical Plant, and Business Base
Raymond Colladay
RC Space Enterprises, Inc.

Fact-finding Visit to GE Aircraft Engines in Evendale, Ohio
Eli Reshotko
Case Western Reserve University

SIXTH MEETING

August 4-5, 2000
Wyndham Colorado Springs
5580 Tech Center
Colorado Springs, Colo.

Writing Session

SEVENTH MEETING

August 23-24, 2000
National Research Council
Green Building
2001 Wisconsin Avenue, N.W.
Washington, D.C.

Space Industrial Base Study
Hendrick Ruck
Air Force Research Laboratory

Air Force Technology Vision
Herbert Carlson
Air Force Office of Scientific Research

Space Industrial Base Study
Erik Anderson
Booz Allen & Hamilton

EIGHTH MEETING

September 8, 2000
National Research Council
Green Building
2001 Wisconsin Avenue, N.W.
Washington, D.C.

Writing Session

NINTH MEETING

October 16-17, 2000
National Research Council
Green Building
2001 Wisconsin Avenue, N.W.
Washington, D.C.

Writing Session

Appendix B

Biographical Sketches of Committee Members

Robert R. Everett (*Chair*), a member of the National Academy of Engineering (NAE), is retired president and chief executive officer (CEO) of the MITRE Corporation; he is currently an honorary trustee and a member of MITRE's board of directors. In 1989, Mr. Everett, a pioneer in the development of digital computers, was awarded the National Medal of Technology by President Bush for his work in real-time computer technologies and applications. Mr. Everett received a B.S. from Duke University and an M.S. from the Massachusetts Institute of Technology. He was a member of the Defense Science Board (DSB) from 1987to 1993 and chair of the DSB from 1988 to 1989. He has served as a senior scientist of the Air Force Scientific Advisory Board and as chairman of the Federal Aviation Administration's Research, Engineering, and Development Committee. He is also a member of the Ballistic Missile Defense Organization's Advisory Committee.

Gordon R. England (*Vice Chair,* September 1999 to May 2001) resigned from his position as vice chair of the study in May 2001 due to his nomination and eventual acceptance as Secretary of the Navy. At that time he was executive vice president of General Dynamics Corporation. Mr. England began his 38-year career as an engineer with Honeywell, where he worked on the Gemini space program. He was also a program manager on the E-2C program with Litton Industries. Mr. England first joined General Dynamics in October 1966 as an avionics design engineer with the company's airplane division in Fort Worth, Texas. After holding various engineering and management positions with the Fort Worth division, he became president and an executive vice president of the corporation in January 1991. When the Fort Worth division was sold to Lockheed in March 1993, he became the president of Lockheed Fort Worth. He retired in 1995 and directed his own consulting business, GRE Consulting Inc., which deals primarily with mergers and acquisitions. He returned to General Dynamics in 1997. Mr. England has a B.S. in electrical engineering from the University of Maryland and an M.B.A. from Texas Christian University.

Kyle T. Alfriend, a member of the NAE, has had a varied career in academia, government, and industry. Currently a professor and head of the Aerospace Engineering Department at Texas A&M University, Dr. Alfriend has previously worked at Lockheed Missiles and Space, GRC International, the Central Intelligent Agency, the Naval Research Laboratory, the National Aeronautics and Space Administration (NASA), Cornell University, and the Naval Postgraduate School in Monterey as a visiting professor. He has been an editor of many magazines and is currently a member of the Aerospace Industries Association, American Astronautical Society, American Society for Engineering Education, and National Space Society and a fellow of the Society for Engineering Science. His areas of interest include formation-flying satellites, characterization of orbit uncertainty, cataloging of space debris, and autonomous rendezvous and docking.

Oliver C. Boileau, Jr., a member of NAE, was president and CEO of the Grumman Corporation, a subsidiary of Northrop-Grumman. He retired in 1995. During Mr. Boileau's career, he held numerous high-level positions, such as president of Boeing Aerospace and president of General Dynamics. He is currently a private consultant to industry and a member of numerous societies and honorary organizations. He has served on the DSB and many other government boards. Mr. Boileau has a BSEE and MSEE from the University of Pennsylvania and an M.S. in industrial management from the Massachusetts Institute of Technology (MIT).

Michael P.C. Carns, General, U.S. Air Force (Ret), was vice chief of staff of the Air Force prior to his retirement. During his career, General Carns served in a variety of operational and management positions and is well versed in the problems facing the defense industry in retaining talented design personnel. He is a member of the DSB, which has recently studied similar issues. General Carns is a graduate of the Air Force Academy and received an M.B.A. from Harvard University.

Raymond S. Colladay retired as president of Lockheed Martin Astronautics. Earlier positions included director of the Defense Research Projects Agency (DARPA) and associate administrator of NASA. Dr. Colladay has served on the Aeronautics and Space Engineering Board of the National Research Council (NRC) and has demonstrated his abilities as chair of the Advanced Space Technology Committee. He has also been a member of the DSB and various other U.S. Department of Defense (DoD) and NASA boards. He is a fellow of AIAA and of the American Astronautical Society. He earned a Ph.D. in mechanical engineering from Michigan State University and an M.B.A. from Harvard University. He is especially knowledgeable about how NASA deals with design infrastructure issues.

John W. Douglass, Brigadier General, U.S. Air Force (Ret), had a distinguished military career in contracting, engineering, and test and evaluation. He also served as the special assistant to the UnderSecretary of Defense for Acquisition and in the White House as director of National Security Programs. After retiring from the Air Force, Mr. Douglass was appointed Assistant Secretary of the Navy for research, development, and acquisition. He is currently president and CEO of the Aerospace Industries Association (AIA). Mr. Douglass earned M.S. degrees from Fairleigh Dickinson University and Texas Technology University.

Robert B. Ormsby, Jr., a member of the NAE, has held many important positions at Lockheed, including group president of Lockheed Aeronautical Systems Group and member of the Lockheed Board of Directors. Mr. Ormsby was chair of the NASA Aeronautics Advisory Committee and is a fellow of AIAA. He graduated from Georgia Institute of Technology with a B.S. in aeronautical engineering and completed the nine-month Sloan Executive Management program at Stanford University. Mr. Ormsby has a strong background in aeronautics and engineering design.

George A. Paulikas is retired executive vice president of the Aerospace Corporation, where his chief responsibility was execution of the launch readiness verification process for boosters and spacecraft. Dr. Paulikas participated in more than 150 space launches and spacecraft operations and was intimately involved in the development and flight of a number of advanced space programs critical to national security. Dr. Paulikas has served as consultant to the NASA Office of Space Sciences and the Lawrence Berkeley Laboratory and the Institute for Defense Analysis. His expertise is in national security space systems.

Thomas M. Perdue is currently senior vice president of the Signature Group, a systems integration firm specializing in architectural design and implementation of large-scale local and wide-area networks and the associated communications systems for enterprise-wide application. Prior to his position with the Signature Group, Mr. Perdue was Principal Assistant Deputy UnderSecretary of Defense (Advanced Technology) with responsibility for the development and oversight of advanced concept technology demonstrations (ACTDs), with particular emphasis on the transition of ACTDs into acquisition and operation use. Mr. Perdue's expertise is in military procurement and contracting.

Winfred M. Phillips is currently associate vice president for the Engineering and Industrial Experiment Station, College of Engineering, University of Florida. He was previously a professor in the Department of Aerospace Engineering, at Pennsylvania State University. Dr. Phillips' expertise is in the maintenance of an educated, trained, innovative force of engineering and science professionals who can support the national aerospace infrastructure.

Herman M. Reininga has been vice president of operations for Rockwell Collins since 1996. He is responsible for manufacturing, material, quality, and facilities and manufacturing activities throughout Rockwell Collins. Mr. Reininga, a recognized leader in integrating commercial and defense manufacturing technology and manufacturing policies, issues, and processes for government and industry. has testified before the Senate Armed Services Committee on Defense Technology, Acquisition, and Industrial Base. From 1990 to 1992, Mr. Reininga was a member of the DSB. Mr. Reininga has an extensive knowledge in manufacturing and industrial engineering and management.

Eli Reshotko, a member of the NAE, is currently Kent H. Smith Professor Emeritus of Engineering at Case Western Reserve University. He is a fellow of the AIAA, American Society of Mechanical Engineers, American Physical Society, and American Academy of Mechanics, of which he is also past president. He is coauthor of more than 100 publications and a member or chair of many task forces, committees, and governing boards. His areas of expertise are viscous effects in external and internal aerodynamics, two-dimensional and three-dimensional compressible boundary layers, heat transfer, stability and transition of viscous flows (both incompressible and compressible), and low-drag technology for aircraft and underwater vehicles.

Michael D. Rich held many important positions at RAND, including vice president, National Security Research. Mr. Rich was the corporate officer in charge of the National Defense Research Institute, the Federally Funded Research and

Development Center that performs research and policy analysis for the Office of the Secretary of Defense and defense agencies, such as the Defense Nuclear Agency and Defense Communications Agency. From 1987 to 1990, he directed two congressionally mandated studies for the undersecretary of defense for acquisition on the B-2 program acquisition strategy. From 1989 to1992, Mr. Rich served on the Executive Committee of the Army STAR study conducted under the auspices of the NRC. Mr. Rich's expertise is in the industrial base.

Harold W. Sorenson has worked at General Dynamics/Astronautics in San Diego, California, and at AC Electronics Division of General Motors in El Segundo, California. After a year as a guest scientist at the Institute for Guidance and Control in Oberpfaffenhofen, West Germany, he joined the faculty at the University of California, San Diego, as professor of engineering sciences. He remained on the faculty until he became group vice president of the Air Force C3I Group at the Mitre Corporation in Bedford, Massachusetts. He later became corporate director for Air Force Systems and Bedford Operations and senior vice president and general manager for Air Force C3 Systems. He was also a member of the Defense Intelligence Agency Scientific Advisory Board. Mr. Sorenson's expertise is in control and information systems.